Electrical Installation
Calculations: Advanced

Electrical Installation Calculations: Advanced

For technical certificate and NVQ level 3

Christopher Kitcher and A.J. Watkins

Eighth Edition

Routledge
Taylor & Francis Group

LONDON AND NEW YORK

Eighth edition published 2014
by Routledge
2 Park Square, Milton Park, Abingdon, Oxon OX14 4RN

and by Routledge
711 Third Avenue, New York, NY 10017

Routledge is an imprint of the Taylor & Francis Group, an informa business

First edition published 1957
Seventh edition published by Elsevier 2009

British Library Cataloguing in Publication Data
A catalogue record for this book is available from the British Library

Library of Congress Cataloging in Publication Data
  Kitcher, Chris.
  Electrical installation calculations : advanced : for technical certificate
  and NVQ level 3 / Christopher Kitcher and A. J. Watkins. — Eighth edition.
    pages cm
  1. Electric wiring—Mathematics.   2. Electric wiring—Standards—Great
  Britain.   3. Electrical engineers—Certification—Great Britain.
  4. National Vocational Qualifications (Great Britain)   I. Watkins, A. J.
  (Albert James), 1925-   II. Title.
  TK3221.K57 2014
  621.319'24—dc23

                                                              2013021029

ISBN: 978-0-415-81003-6 (pbk)
ISBN: 978-1-315-86312-2 (ebk)

Typeset in Rotis Sans Serif
by RefineCatch Limited, Bungay, Suffolk

Printed and bound in Great Britain by
TJ International Ltd, Padstow, Cornwall

MIX
Paper from
responsible sources
FSC® C013056
www.fsc.org

Table of Contents

Preface

Being able to carry out mathematical calculations is a vital part of electrical installation courses and indeed electrical installation work.

The structure of electrical installation courses continually changes as do the course titles and numbers, however electrical science remains the same, and like it or not anyone wanting to become an electrician will need to have a good understanding of how to carry out electrical calculations.

The calculations which need to be performed vary from those which an electrician needs almost on a daily basis, such as cable calculation or the amount of energy required to run a particular piece of equipment, to more complex calculations such as those required for electromagnetism.

This book will show you how to carry out these calculations as simply as possible using electronic calculator methods. These methods will be useful both in the classroom and the workplace. It is not necessary for you to have a deep understanding of how the mathematical functions are performed. Each topic is shown using a step-by-step process with lots of exercises provided to give you the opportunity to test yourself at the end of each chapter.

This edition has been completely updated to the 17th edition of BS 7671 amendment 1: 2011 and the *IET On-Site Guide*, useful references are made to these documents throughout.

It does not matter which electrical course you are attending, this book along with the basic calculations book will be invaluable.

Use of Calculators

Throughout books 1 and 2 the use of a calculator is encouraged. Your calculator is a tool, and like any tool practice is required to perfect their use. A scientific calculator will be required, and although they differ in the way the functions are carried out, the end result is the same.

The examples are given using a Casio fx-83MS.

The character printed on the button is the function performed when the button is pressed. To use the small letter functions on the top of any button the **shift** button must be used.

Practice is important.

Syntax error: will appear when the figures are entered in the wrong order.

x^2: will multiply a number by itself, eg $6 \times 6 = 36$. On the calculator this would be $6 \ x^2 = 36$. When a number is multiplied by itself it is said to be **squared**.

x2: will multiply a number by itself and then the total by itself again, eg when we enter 4 on calculator $x^3 = 64$. When a number is multiplied in this way it is said to be **cubed**.

$\sqrt{}$: will give you the number which achieves your total by being multiplied by itself, eg $\sqrt{36} = 6$. This is said to be the **square root** of a number, and is the opposite of **squared**.

$\sqrt[3]{}$: will give you the number which when multiplied by itself three times will be your total $\sqrt[3]{64} = 4$. This is said to be the **cube root**.

x^{-1}: will divide 1 by a number, eg $\frac{1}{4} = 0.25$. This is the *reciprocal* button and is useful in this book for finding the resistance of resistors in parallel and capacitors in series.

EXP: is for the powers of 10 function, eg $25 \times 1000 = 25$ EXP $\times 10^3 = 25\ 000$.

Enter into your calculator 25 EXP 3 = 25000. (Do not enter the $\times$ or the number 10.)

If a calculation shows 10^{-3}, eg 25×10^{-3}, enter 25 EXP –3 = (0.025) (*when using EXP if a minus is required use the button (–)*).

Brackets: these should be used to carry out a calculation within a calculation.

Example calculation $\dfrac{32}{(0.8 \times 0.65 \times 0.94)} = 65.4$

Enter into calculator $32 \div (0.8 \times 0.65 \times 0.94 = 65.46)$

Remember **Practice makes Perfect.**

Simple Transposition of Formulae

To find an unknown value:

■ the subject must be on the top line and must be on its own;

■ the answer will always be on the top line;

■ to get the subject on its own, values must be moved;

■ any value that moves across the = sign must move from above the line to below the line; or

■ from below the line to above the line.

EXAMPLE 1

$3 \times 4 = 2 \times 6$

$3 \times 4 = 2 \times ?$

Transpose to find **?**

$$\frac{3 \times 4}{2} = 6$$

EXAMPLE 2

$$\frac{2 \times 6}{?} = 4$$

Step 1 $2 \times 6 = 4 \times ?$

Step 2 $\dfrac{2 \times 6}{?} = 4$

Answer $\dfrac{2 \times 6}{4} = 3$

EXAMPLE 3

$5 \times 8 \times 6 = 3 \times 20 \times$?

Step 1 move 3×20 away from unknown value, as the known values move across the = sign they must move to bottom of equation $\dfrac{5 \times 8 \times 4}{3 \times 20} = ?$

Step 2 Carry out the calculation

$\dfrac{5 \times 8 \times 6}{3 \times 20} = \dfrac{240}{60} = 4$

Therefore

$5 \times 8 \times 6 = 240$ *or* $3 \times 20 \times 4 = 240$ or $5 \times 8 \times 6 = 3 \times 20 \times 4$

SI Units

In the United Kingdom and the rest of Europe the units for measuring different properties are known as SI units.

SI stands for **Système Internationale**.

All units are derived from seven base units.

Base quantity	Base unit	Symbol
Time	Second	s
Electrical current	Ampere	A
Length	Metre	m
Mass	Kilogramme	kg
Temperature	Kelvin	K
Luminous intensity	Candela	cd
Amount of substance	Mole	mol

SI DERIVED UNITS

Derived quantity	Name	Symbol
Frequency	hertz	Hz
Force	Newton	N
Energy, work, quantity of heat	joule	J
Electric charge, quantity of electricity	coulomb	C
Power	watt	W
Potential difference, electromotive force	volt	V or U
Capacitance	farad	F
Electrical resistance	ohm	Ω
Magnetic flux	weber	Wb
Magnetic flux density	tesla	T
Inductance	henry	H
Luminous flux	lumen	cd
Area	square metre	m^2
Volume	cubic metre	m^3
Velocity, speed	metre per second	m/s
Mass density	kilogramme per cubic metre	kg/m^3
Luminance	candela per square metre	cd/m^2

SI UNIT PREFIXES

Name	Multiplier	Prefix	Power of 10
Tera	1000 000 000 000	T	1×10^{12}
Giga	1000 000 000	G	1×10^{9}
Mega	1000 000	M	1×10^{6}
Kilo	1000	k	1×10^{3}
Unit	1		
milli	0.001	m	1×10^{-3}
micro	0.000 001	μ	1×10^{-6}
nano	0.000 000 001	η	1×10^{-9}
pico	0.000 000 000 001	ρ	1×10^{-12}

EXAMPLES

mA   milliamp = one thousandth of an ampere

km   kilometre = one thousand metres

μv   micro volt = one millionth of a volt

GW   Giga watt = one thousand million watts

kW   Kilowatt = one thousand watts

Calculator example:

1 kilometre is 1 metre $\times 10^3$

Enter into calculator 1 EXP 3 = (*1000*) metres

1000 metres is 1 kilometre $\times 10^{-3}$

Enter into calculator 1000 EXP −3 = (*1*) kilometre

1 micro volt is 1 volt × 10^{-6}

Enter into calculator 1 EXP −6 = (1^{-06} or 0.000001) volts

(note 6th decimal place)

Conductor Colour Identification

	Old colour	New colour	Marking
Phase 1 of a.c.	Red	Brown	L1
Phase 2 of a.c.	Yellow	Black	L2
Phase 3 of a.c.	Blue	Grey	L3
Neutral of a.c.	Black	Blue	N

Note: Great care must be taken when working on installations containing old and new colours. Although we still have three phases they are now referred to as lines 1, 2 and 3.

Alternating Current Circuit Calculations

IMPEDANCE

In d.c. circuits the current is limited by resistance. In a.c. circuits the current is limited by impedance *(Z)*. Resistance and impedance are measured in ohms.

For this calculation ohms law is used and *Z* is substituted for *R*.

$\dfrac{U}{Z} = I$ or voltage (U) ÷ impedance (ohms) = current (amperes)

Figure 1 Voltage, impedance, current triangle

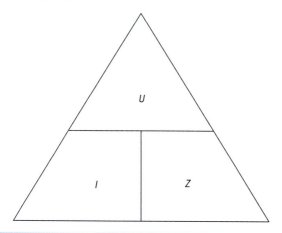

EXAMPLE 1

The voltage applied to a circuit with an impedance of 6Ω is 200V.

Calculate the current in the circuit.

$$\frac{U}{Z} = I$$

$$\frac{200}{6} = 33.33\,A$$

EXAMPLE 2

The current in a 230V single-phase motor is 7.6A. Calculate the impedance of the circuit.

$$\frac{U}{I} = Z$$

$$\frac{230}{7.6} = 30.26\,Ω$$

EXAMPLE 3

A discharge lamp has an impedance of 275Ω and the current drawn by the lamp is 0.4A. Calculate the voltage.

$$Z \times I = U$$

$$275 \times 0.4 = 110\ volts$$

EXAMPLE 4

The current through an impedance of 32Ω is 8A. Calculate the voltage drop across the impedance.

$$U = I \times Z$$

$$= 8 \times 32$$

$$= 256V$$

EXAMPLE 5

The current through an electric motor is 6.8A at 230V. Calculate the impedance of the motor.

$U = IZ$

(transpose for Z) $Z = \dfrac{U}{I}$

$= \dfrac{230}{6.8}$

$= 33.82 \ \Omega$

EXAMPLE 6

An a.c. coil has an impedance of 430Ω. Calculate the voltage if the coil draws a current of 0.93A.

$U = I \times Z$

$= 0.93 \times 430$

$= 400\text{V}$

EXERCISE 1

1 Complete the following table:

Volts (a.c.)			230	400	100	25	230	
Current (A)	0.1	15	0.5		0.01		180	25
Impedance (Ω)	100	15		1000		0.05		25

2 Complete the following table:

Current (A)	1.92	3.84	18.2		7.35	4.08		8.97
Volts (a.c.)		7.5		230	107		400	235
Impedance (Ω)	2.45		12.4	96.3		56	96	

3 Complete the following table:

Impedance (Ω)	232	850		0.125		1050	129	
Volts (a.c.)		230	400	26.5	0.194		238	245
Current (A)	0.76		0.575		0.0065	0.436		0.056

4 A mercury vapour lamp takes 2.34A when the mains voltage is 23V. Calculate the impedance of the lamp circuit.

5 An inductor has an impedance of 365Ω. How much current will flow when it is connected to a 400V a.c. supply?

6 A coil of wire passes a current of 55A when connected to a 120V d.c. supply but only 24.5A when connected to a 110V a.c. supply. Calculate (a) the resistance of the coil, (b) its impedance.

7 Tests to measure the impedance of an earth fault loop were made in accordance with BS 7671 and the results for five different installations are given below. For each case, calculate the value of the loop impedance.

	Test voltage, a.c. (V)	Current (A)
(a)	9.25	19.6
(b)	12.6	3.29
(c)	7.65	23.8
(d)	14.2	1.09
(e)	8.72	21.1

8 The choke in a certain fluorescent-luminaire fitting causes a voltage drop of 150V when the current through it is 1.78A. Calculate the impedance of the choke.

9 Complete the following table:

Volts (a.c.)	61.1		153	193	
Current (A)	2.3	4.2		7.35	9.2
Impedance (Ω)		25	25		25

Plot a graph showing the relationship between current and voltage. From the graph, state the value of the current when the voltage is 240V.

10 The alternating voltage applied to a circuit is 230V and the current flowing is 0.125A. The impedance of the circuit is:

(a) 5.4Ω   (b) 1840Ω   (c) 3.5Ω   (d) 184Ω

11 An alternating current of 2.4A flowing in a circuit of impedance 0.18Ω produces a voltage drop of

(a) 0.075V   (b) 13.3V   (c) 0.432V   (d) 4.32V

12 When an alternating e.m.f. of 150V is applied to a circuit of impedance 265Ω, the current is

(a) 39 750A   (b) 1.77A   (c) 5.66A   (d) 0.566A

INDUCTIVE REACTANCE

When an a.c. current is passed through a conductor a magnetic field is created around the conductor. If the conductor is wound into a coil the magnetic field is increased. Where there are significant magnetic fields in a circuit there is opposition to the flow of current, this opposition is called *inductive reactance*. The opposition caused by inductive reactance is in addition to the opposition caused by the resistance caused by the wires.

In this section we will assume that the resistance of the circuits is so low that it may be ignored and that the only opposition to the flow of current is that caused by the inductive reactance.

The formulae for inductive reactance is $X_L = 2\pi f L$ (answer in ohms).

L is the inductance of the circuit or coil of wire and is stated in *henrys* (H).

f is the frequency of the supply in hertz (Hz) (Figure 2).

Figure 2 An inductive circuit

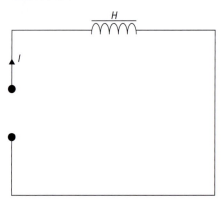

EXAMPLE 1

Calculate the inductive reactance of a coil which has an inductance of 0.03H when connected to a 50Hz supply.

$X_L = 2\pi fL$

$= 2 \times 3.142 \times 50 \times 0.03 = 9.42\Omega$

EXAMPLE 2

Calculate the inductive reactance of the coil in Example 1 when connected to a 60Hz supply.

$X_L = 2\pi fL$

$= 2 \times 3.142 \times 60 \times 0.03 = 11.31\Omega$

It can be seen from this calculation that if the frequency increases the inductive reactance will also increase.

EXAMPLE 3

An inductor is required to cause a voltage drop of 180V when a current of 1.5A is passed through it at a frequency of 50Hz. Calculate the value of the inductor:

$U_L = I \times X_L$ (this is ohms law with inductive reactance instead of resistance)

Transpose $\dfrac{U}{I} = X_L$ $\dfrac{180}{1.5} = 120\Omega$

$X_L = 2\pi fL$

$120 = 2 \times 3.142 \times 50 \times L$

Transpose $\dfrac{120}{2 \times 3.142 \times 50} = 0.381H$

On calculator enter $120 \div (2\pi \times 50) =$ (answer 0.382H)

EXERCISE 2

1 Calculate the inductive reactance of a coil having an inductance of 0.015H when a 50Hz current flows in it.
2 A coil is required to have an inductive reactance of 150Ω on a 50Hz supply. Determine its inductance.
3 Complete the following table:

Inductance (H)	0.04		0.12	0.008	
Frequency (Hz)	50	50			60
Reactance (Ω)		50	36	4.5	57

4 A coil of negligible resistance causes a voltage drop of 98V when the current through it is 2.4A at 50Hz. Calculate (a) the reactance of the coil, (b) its inductance.
5 A reactor allows a current of 15A to flow from a 230V 50Hz supply. Determine the current which will flow at the same voltage if the frequency changes to (a) 45Hz, (b) 55Hz. Ignore the resistance.
6 Calculate the inductive reactance of coils having the following values of inductance when the supply frequency is 50Hz.

   (a) 0.012H

   (b) 0.007H

(c) 0.45mH

(d) 350µH

(e) 0.045H

7 Determine the inductances of the coils which will have the following reactance to a 50Hz supply:

(a) 300Ω

(b) 25Ω

(c) 14.5Ω

(d) 125Ω

(e) 5Ω

8 Calculate the voltage drop across a 0.24H inductor of negligible resistance when it carries 5.5A at 48Hz.

9 An inductor of 0.125H is connected to an a.c. supply at 50Hz. Its inductive reactance is

(a) 39.3Ω   (b) 0.79Ω   (c) 0.025Ω   (d) 393Ω

10 The value in *henrys* of an inductor which has an inductive reactance of 500Ω when connected in an a.c. circuit at frequency 450Hz is

(a) 1.77H

(b) 14×10^6H

(c) 0.177H

(d) 0.071×10^{-6}H

CAPACITIVE REACTANCE

When a capacitor is connected to an a.c. supply, the current flow is limited by the *reactance* of the capacitor (X_C).

Formulae for capacitive reactance $X_C = \dfrac{10^6}{2\pi f C}$

Figure 3 Capacitive reactance

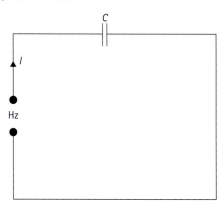

C is the capacitance of the capacitor measured in microfarads (μF).

f is the frequency of the supply in hertz (Hz).

(It should be noted that d.c. current will not flow with a capacitor in the circuit, it will simply charge and then stop) (Figure 3).

EXAMPLE 1

Calculate the reactance of a 70μF capacitor to a 50Hz supply:

$$X_C = \frac{10^6}{2\pi f C}$$

$$\frac{10^6}{2 \times 3.142 \times 50 \times 70} = 45.47\Omega$$

Enter on calculator EXP 6 ÷ (2π × 50 × 70) =

(answer 45.47).

EXAMPLE 2

A power factor improvement capacitor is required to take a current of 7.5A from a 230V 50Hz supply. Determine the value of the capacitor.

For this calculation ohms law is used and R is substituted by X_c

Step 1

$U_c = I \times X_c$

$230 = 7.5 \times X_c$

Transpose for X_c

$\dfrac{230}{7.5} = X_c$

$\dfrac{230}{7.5} = 30.6\Omega$

Step 2 to find C

$X_c = \dfrac{10^6}{2\pi f C} \; transpose \, C = \dfrac{10^6}{2\pi \times f \times X_c}$

$C = \dfrac{10^6}{(2 \times 3.142 \times 50 \times 30.6)} = 104 \; answer \; is \; in \; micro \, f \, arads \; (\mu F)$

Note: Simply change places of X_c and C.

Enter on calculator *EXP* 6 ÷ (2π × 50 × 30.6) *or*
EXP 6 ÷ (2 × 3.142 × 50 × 30.16)

Note: The EXP button is sometimes shown on a calculator as × 10^x.

EXERCISE 3

1 Determine the reactance of each of the following capacitors to a 50Hz supply. (Values are all in microfarads.)

    (a) 60

    (b) 25

    (c) 40

    (d) 150

    (e) 8

    (f) 12

(g) 250

(h) 95

(i) 16

(j) 75

2 Calculate the value of capacitors which have the following reactances at 50Hz. (Values are all in ohms).

(a) 240

(b) 75

(c) 12

(d) 4.5

(e) 36

(f) 16

(g) 45

(h) 400

(i) 30

(j) 72

3 Calculate the value of a capacitor which will take a current of 25A from a 230V 50Hz supply.

4 A capacitor in a certain circuit is passing a current of 0.2A and the voltage drop across it is 100V. Determine its value in microfarads. The frequency is 50Hz.

5 Calculate the voltage drop across a 5μF capacitor when a current of 0.25A at 50Hz flows through it.

6 In order to improve the power factor of a certain installation, a capacitor which will take 15A from the 230V supply is required. The frequency is 50Hz. Calculate the value of the capacitor.

7 In one type of twin-tube fluorescent fitting, a capacitor is connected in series with one of the tubes. If the value of the capacitor is 7μF, the current through it is 0.8A, and the supply is at 50Hz, determine the voltage across the capacitor.

8 A machine designed to work on a frequency of 60Hz has a power-factor-improvement capacitor which takes 12A from a 110V supply. Calculate the current the capacitor will take from the 110V 50Hz supply.

9 A capacitor takes a current of 16A from a 400V supply at 50Hz. What current will it take if the voltage falls to 380V at the same frequency?

10 A 22μF capacitor is connected in an a.c. circuit at 50Hz. Its reactance is

(a) 0.000145Ω   (b) 6912Ω   (c) 6 912 000Ω   (d) 145Ω

11 The value in microfarads of a capacitor which has a capacitive reactance of 100Ω when connected to a circuit at 50Hz is

(a) 31.8μF   (b) 318μF   (c) 0.0000318μF   (d) 0.0314μF

IMPEDANCE IN SERIES CIRCUITS

When resistance (R) is in a circuit with reactance (X_L or X_C) the combined effect is called impedance (Z); this is measured in ohms.

For series circuits the calculation for impedance (Z) is

$Z^2 = R^2 + X^2$ or $Z = \sqrt{R^2 + X^2}$

In this calculation X is f or X_C or X_L

Where the circuit contains inductive reactance (X_C) and capacitive reactance (X_L)

$X = X_C - X_L$ or $X_L - X_C$

X will be the largest reactance minus the smallest reactance.

An inductor coil will always possess both inductance (the magnetic part of the circuit) and resistance (the resistance of the wire), together they produce impedance. Although inductance and impedance cannot be physically separated, it is convenient for the purpose of calculation to show them separately in a circuit diagram (Figure 4).

Figure 4 An inductor and resistor

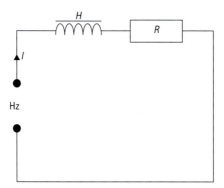

EXAMPLE 1

A coil has a resistance of 6Ω and an inductance of 0.09H. Calculate its impedance to a 50Hz supply.

Step 1. Inductive reactance $X_L = 2\pi fL$

$2\pi \times f \times 0.09$

$2 \times 3.142 \times 50 \times 0.09 = 28.27\Omega$

Note: A common error is to add the resistance and inductance treating it as a
d.c. circuit.

Step 2. $Z^2 = R^2 + X_L^2$

or $Z = \sqrt{R^2 + X^2}$

$Z = \sqrt{6^2 + 28.27^2}$

28.9Ω

Enter into calculator 28.9Ω $6X^2 + 28.27X^2 = \sqrt{}$ = $(Ans)28.9\Omega$

EXAMPLE 2

A coil passes a current of 23A when connected to a 230V d.c. supply but only 8A when connected to a 230V supply.

When connected to a d.c. circuit the coil's resistance is only that of the wire in the coil, this can be calculated using ohms law.

On D.C. $U = I \times R$

$$\frac{U}{I} = R$$

$$\frac{230}{23} = 10\Omega (resistance)$$

On an a.c. circuit, reactance will be produced, as this is an inductive circuit it will be inductive reactance (X_L).

The combined effect of the resistance and reactance of the coil is the impedance (Z).

Step 1. On a.c. $U = I \times Z$

$230 = 8 \times Z$

Transpose $\frac{230}{8} = 28.75\Omega \ impedance \ (Z)$

Step 2. To find the inductance of the coil.

$Z^2 = R^2 + X_L^2$

$X_L^2 = Z^2 - R^2$

$X_L = \sqrt{28.7^2 - 10^2}$

$X_L = 26.44\Omega$

Enter on calculator 28.27 $X^2 - 10^2 = \sqrt{} = $ (Answer) 26.44Ω

Step 3. $X_L = 2\pi fL$

$26.44 = 2 \times 3.142 \times 50 \times L$

Transpose $\frac{26.44}{(2 \times 3.142 \times 50)} = L = 0.084H$

Enter on calculator 26.44 $\div$ (2 $\times$ 3.142 $\times$ x 50) = Ans

Figure 5 Circuit in example 3

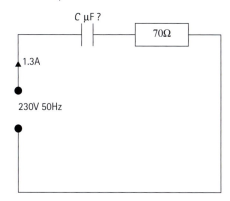

EXAMPLE 3

A 70Ω resistor is wired in series with a capacitor of an unknown value to a 230V 50Hz supply. Calculate the value of the capacitor in microfarads if a current of 1.3A flows (Figure 5).

First find impedance of circuit (Z)

Step 1. $U = I \times Z$

$230 = 1.3 \times Z$

$Z = \dfrac{230}{1.3}$

$Z = 176.92\Omega$

Step 2. Next find capacitive reactance X_C

$Z^2 = R^2 + X_C^2$

$176.92 = \sqrt{70^2 + X_C^2}$

Transpose *for* X_C

$X_C = \sqrt{176.92^2 - 70^2}$

$X_C = 162.48\Omega$

Figure 6 Finding the capacitance

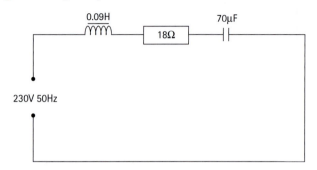

Now find capacitance

Step 3. $X_c = \dfrac{10^6}{2\pi f C}$

Transpose for C

$C = \dfrac{10^6}{2\pi f \, X_L}$

$C = \dfrac{10^6}{2 \times 3.142 \times 50 \times 162.48}$

19.59 µF is the capacitor value

Enter on calculator *EXP or X^{-1}*6 ÷ (2 × 3.142 × 50 × 162.48 = (*Ans*)

EXAMPLE 4

A coil of inductance of 0.09H and a resistance of 18Ω is wired in series with a 70µF capacitor to a 230V 50Hz supply. Calculate the current which flows and the voltage drop across the capacitor.

The first step is to calculate inductive and capacitive reactance.

Inductive reactance $X_L = 2\pi f L$

$$= 2 \times 3.142 \times 50 \times 0.09$$

$$= 28.27\Omega$$

Capacitive reactance $X_c = \dfrac{10^6}{2\pi f C}$

$$= \dfrac{10^6}{2 \times 3.142 \times 50 \times 70}$$

$= 45.47\Omega$

Enter on calculator *EXP or X^{-1}*6 × 2 × 3.142 × 50 × 70 = (*Ans*)

The second step is to find the actual reactance for the circuit which is the largest reactance minus the smallest reactance.

For this circuit $X = X_C - X_L$

$$= 45.47 - 28.27$$

$= 17.2\Omega$ (this is X_C as the capacitive reactance is larger than the inductive reactance)

The third step is to calculate the impedance for the circuit (Z)

Impedance Z is found

$Z^2 = R^2 + X_C^2$

$Z^2 = 18^2 + 17.2^2$

$Z = \sqrt{18^2 + 17.2^2}$

Enter on calculator 18 X^2 + 17.2X^2 = $\sqrt{}$ = (Ans)

$Z = 24.88\Omega$

The fourth step is to calculate current (I)

$U = I \times Z$

$230 = I \times 24.88$

Transpose for I

$\dfrac{230}{24.88} = 9.24A$

As this current is common to the whole circuit, the voltage across the capacitor and the inductor can be calculated.

Figure 7 Calculating the voltage

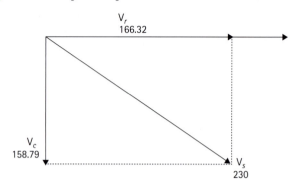

If a phasor is required the current is the reference conductor.

Voltage across capacitor $U_c = I \times X_c$

$9.24 \times 45.46 = 420$ volts

Voltage across inductor $U_l = I \times X_L$

$9.24 \times 28.27 = 261\ 21$ volts

Note: Both voltages are higher than the 230V supply. This often happens in a.c. circuits. The voltages do not add up as in d.c. circuits.

EXERCISE 4

1 Complete the following table:

R	15	25	3.64				76.4	0.54
R^2				2250	18.7	40		

2 Complete the following table:

X	29.8		0.16			897		
X^2		0.46		0.9	0.16		54637	0.036

3 A coil of wire has resistance of 8Ω and inductance of 0.04H. It is connected to a supply of 100V at 50Hz. Calculate the current which flows.

4 An inductor of inductance 0.075H and resistance 12Ω is connected to a 230V supply at 50Hz. Calculate the current which flows.

5 Complete the following table:

R (Ω)	14.5		9.63	3.5	57.6	
X (Ω)	22.8	74.6		34.7		49.6
Z (Ω)		159	18.4		4050	107

6 A capacitor of 16μF and a resistor of 120Ω are connected in series. Calculate the impedance of the circuit.

7 A resistor of 200Ω and a capacitor of unknown value are connected to a 230V supply at 50Hz and a current of 0.85A flows. Calculate the value of the capacitor in microfarads.

8 When a certain coil is connected to a 110V d.c. supply, a current of 6.5A flows. When the coil is connected to a 110V 50Hz a.c. supply, only 1.5A flows. Calculate (a) the resistance of the coil, (b) its impedance and (c) its reactance.

9 The inductor connected in series with a mercury vapour lamp has resistance of 2.4Ω and impedance of 41Ω. Calculate the inductance of the inductor and the voltage drop across it when the total current in the circuit is 2.8A. (Assume the supply frequency is 50Hz.)

10 An inductor takes 8A when connected to a d.c. supply at 230V. If the inductor is connected to an a.c. supply at 230V 50Hz, the current is 4.8A. Calculate (a) the resistance, (b) the inductance and (c) the impedance of the inductor.

11 What is the function of an inductor in an a.c. circuit? When a d.c. supply at 230V is applied to the ends of a certain inductor coil, the current in the coil is 20A. If an a.c. supply at 230V 50Hz is applied to the coil, the current in the coil is 12.15A. Calculate the impedance, reactance, inductance, and resistance of the coil. What would be the general effect on the current if the frequency of the a.c. supply were increased?

12 The inductor in a discharge lighting circuit causes a voltage drop of 120V when the current through it is 2.6A. Determine the size in microfarads of a capacitor which will produce the same voltage drop at the same current value. (Neglect the resistance of the inductor. Assume the supply frequency is 50Hz.)

13 A circuit is made up of an inductor, a resistor and a capacitor all wired in series. When the circuit is connected to a 50Hz a.c. supply, a current of 2.2A

flows. A voltmeter connected to each of the components in turn indicates 220V across the inductor, 200V across the resistor and 180V across the capacitor. Calculate the inductance of the inductor and the capacitance of the capacitor. At what frequency would these two components have the same reactance value? (Neglect the resistance of the inductor.)

14 What are meant by the following terms used in connection with alternating current: resistance, impedance and reactance? A voltage of 230V, at a frequency of 50Hz, is applied to the ends of a circuit containing a resistor of 5Ω, an inductor of 0.02H, and a capacitor of 150μF, all in series. Calculate the current in the circuit.

15 A coil of resistance 20Ω and inductance 0.08H is connected to a supply at 240V 50Hz. Calculate (a) the current in the circuit, (b) the value of a capacitor to be put in series with the coil so that the current shall be 12A.

16 For the circuit shown in Figure 8, the voltage V is

    (a) 94V   (b) 14V   (c) 10V   (d) 0.043V

Figure 8 Circuit for question 16

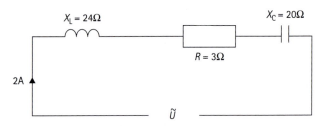

17 An inductor has inductance 0.12H and resistance 100Ω. When it is connected to a 100V supply at 150Hz, the current through it is

    (a) 1.51A   (b) 0.47A   (c) 0.66A   (d) 0.211A

IMPEDANCE TRIANGLES AND POWER TRIANGLES

For a right-angled triangle (Figure 9), the theorem of Pythagoras states that

$$a^2 = b^2 + c^2$$

Figure 9 Pythagoras theorum

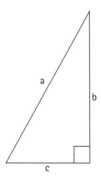

As the relationship between impedance, resistance and reactance in a series circuit is given by an equation of a similar form, $Z^2 = R^2 + X^2$, conditions in such circuits can conveniently be represented by right-angled triangles. In Figure 10, $Z^2 = R^2 + X^2$

where $X = X_L$ (Figure 10(a)) or X_C (Figure 10(b)) and Θ = the *phase angle* of the circuit

$$\sin \phi = \frac{X}{Z} \quad \cos \phi = \frac{R}{Z} \quad \text{and } \tan \phi = \frac{X}{R}$$

$\cos \phi$ is the p.f. of the circuit (Figure 10).

Figure 10 a) inductive reactance b) capacitive reactance

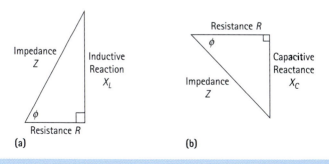

(a)　　　　　　　　　　　　　　　　　(b)

Figure 11 Active and reactive components

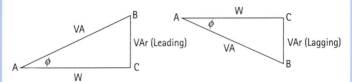

A right-angled triangle is also used to represent the apparent power in a circuit and its active and reactive components (Figure 11).

AB is the product of voltage and current in the circuit (VA).
AC is the true power – the working component (W).
BC is the reactive or wattless component (VAr).
And cos Θ is the power factor (p.f.).
In power circuits, the following multiples of units are used: kVA, kW and kVAr.

EXAMPLE 1

Find Z in Figure 12.

$$Z^2 = R^2 + X^2_L$$
$$= 56^2 + 78^2$$
$$= 3136 + 6084$$
$$= 9220$$
$$\therefore Z = \sqrt{9220}$$
$$= 96.02$$
$$= 96\,\Omega \qquad \text{(correct to three significant figures)}$$

Figure 12 Find Z

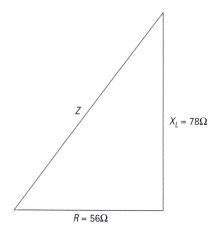

$X_L = 78\Omega$

$R = 56\Omega$

EXAMPLE 2

Find X_C in Figure 13.

$Z^2 = R^2 + X^2_C$

$125^2 = 67.2^2 + X^2_C$

$\therefore X^2_C = 125^2 - 67.6^2$

$\qquad = 15625 - 4570$

$\qquad = 11055$

$\therefore X_C \sqrt{11055} = 105.1$

$\qquad\qquad = 105\Omega$

Alternatively,

$Z^2 = R^2 + X^2_C$

$125^2 = 67.2^2 + X^2_C$

$\therefore X^2_C = 125^2 - 67.6^2$

$\qquad = (125 + 67.6)\,(125 - 67.6)$

$\qquad = 192.6 \times 57.4$

$\qquad = 11055$

$\therefore X_C = \sqrt{11055}$

$\qquad = 105\Omega$

Figure 13 Find X_C

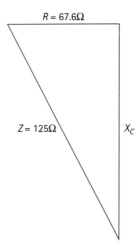

$R = 67.6\Omega$

$Z = 125\Omega$

X_C

EXAMPLE 3

Find Θ in Figure 14.

$$\tan \Theta = \frac{X_L}{R}$$

$$= \frac{15}{20}$$

$$= 0.75$$

$$\therefore \Theta = 36°52'$$

EXAMPLE 4

Find X_C in Figure 15.

$$X_{C/Z} = \sin \Theta$$

$$\frac{X_C}{90} = \sin 48° = 0.7431$$

$$\therefore X_C = 90 \times 0.7431$$

$$= 66.9 \qquad \text{(to three significant figures)}$$

Figure 14 Find Θ

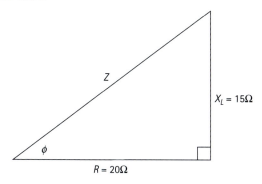

Figure 15 Find X_C

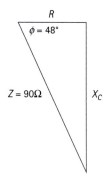

EXAMPLE 5

Find the kVA and kVAr in Figure 16

$$\frac{kW}{kVA} = \_\cos\Theta$$

$$\frac{15}{kVA} = \cos 42° = 0.7431$$

$$\therefore \frac{KVA}{15} = \frac{1}{0.7431}$$

$$kVA = \frac{15}{0.7431}$$

$$= 20.2$$

$$\frac{kVAr}{kW} = \tan\Theta$$

$$\therefore \frac{kVAr}{15} = \tan 42° = 0.9004$$

$$\therefore kVAr = 15 \times 0.9004$$
$$= 13.5kVAr$$

Figure 16 Find kVA and kVAr

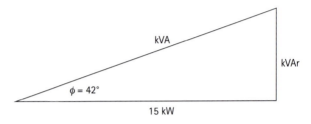

$\phi = 42°$

kVA

kVAr

15 kW

EXAMPLE 6

A coil of 0.2H inductance and negligible resistance is connected in series with a 50Ω resistor to the 230V 50Hz mains (Figure 17). Calculate (a) the current which flows, (b) the power factor, (c) the phase angle between the current and the applied voltage.

Coil reactance $X_L = 2\pi fL$
$$= 2\pi \times 50 \times 0.2$$
$$= 314 \times 0.2$$
$$= 62.8\Omega$$

To find the impedance (Figure 18):

$Z^2 = R^2 + X^2_L$
$$= 50^2 + 62.8^2$$
$$= 2500 + 3944$$
$$= 6444$$
$\therefore Z = \sqrt{6444}$
$$= 80.27\Omega$$

Figure 17 Circuit for example 6

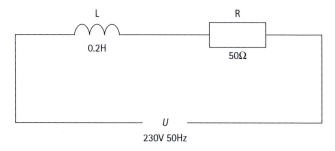

Figure 18 Finding the impedance

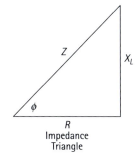

Impedance
Triangle

(a) To find the current,

$U = I \times Z$

$\therefore 230 = I \times 80.27$

$\therefore I = \dfrac{230}{80.27}$

$\quad = 2.86\ A$

(b) Power factor $= \cos\Theta = \dfrac{R}{Z}$

$\quad = \dfrac{50}{80.27}$

$\quad = 0.623\ lag$

(c) The phase angle is the angle whose cosine is 0.623,

$\therefore\ \theta = 51°28'$

EXERCISE 5

1 Find Z in Figure 19.

Figure 19 Find Z

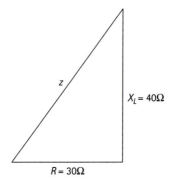

z

$X_L = 40\Omega$

$R = 30\Omega$

2 Find Z in Figure 20.

Figure 20 Find Z

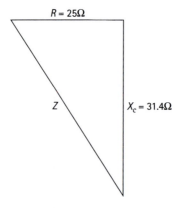

$R = 25\Omega$

Z

$X_c = 31.4\Omega$

3 Find *R* in Figure 21.

Figure 21 Find R

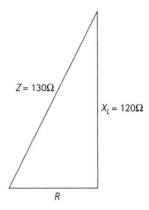

$Z = 130\Omega$

$X_L = 120\Omega$

R

4 Find X_c in Figure 22.

Figure 22 Find X$_c$

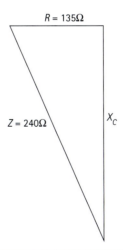

$R = 135\Omega$

$Z = 240\Omega$

X_c

5 Find *R* in Figure 23.

Figure 23 Find R

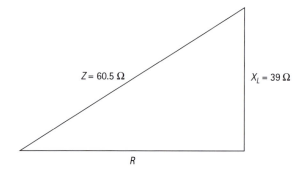

$Z = 60.5\ \Omega$ $X_L = 39\ \Omega$ R

6 Find *Z* in Figure 24.

Figure 24 Find Z

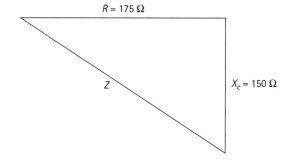

$R = 175\ \Omega$ $X_c = 150\ \Omega$ Z

7 Find *R* in Figure 25.

Figure 25 Find R

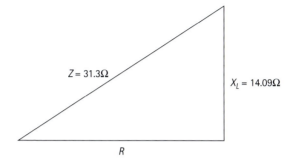

8 Find X_L in Figure 26.

Figure 26 Find X$_L$

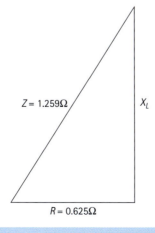

9 Find Z in Figure 27.

Figure 27 Find Z

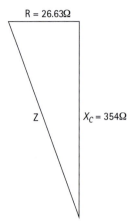

$R = 26.63\Omega$

Z

$X_C = 354\Omega$

10. Find X_L in Figure 28.

Figure 28 Find X_L

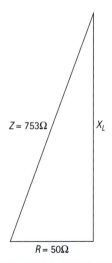

$Z = 753\Omega$

X_L

$R = 50\Omega$

11 Find *R* in Figure 29.

Figure 29 Find R

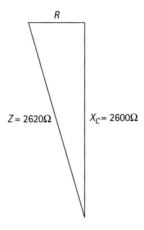

12 Consider the answers to Questions 9 to 11 and then write down the approximate impedance of a coil which has resistance 32Ω and reactance 500Ω.

13 Complete the following table:

Angle φ	30°	45°	60°	90°	52°24′	26°42′	83°12′	5°36′
Sin φ								
Cos φ								
Tan φ								

14 Complete the following table:

Angle φ	33°3′	75°21′	17°15′	64°29′	27°56′	41°53′
Sin φ						
Cos φ						
Tan φ						

15 Complete the following table:

Angle φ								
Sin φ			0.91	0.6			0.9088	
Cos φ		0.9003			0.8			0.4754
Tan φ	0.4000					1.2088		

16 Complete the following table:

Angle φ				38°34′				
Sin φ	0.9661							
Cos φ		0.4341			0.8692	0.3.2.		0.318
Tan φ			0.0950				3.15	

17 Find R and X_L in Figure 30.

Figure 30 Find R and X_L

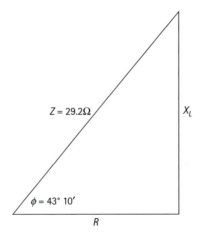

$Z = 29.2\,\Omega$

X_L

$\phi = 43°\ 10′$

R

18 Find R and X_C in Figure 31.

Figure 31 Find R and X$_C$

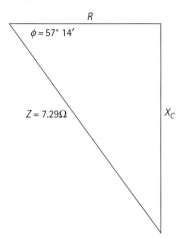

19 Find ϕ in Figure 32.

Figure 32 Find ϕ

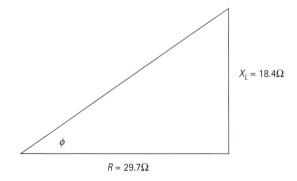

20 Calculate Z and X_L in Figure 33.

Figure 33 Find Z and X$_L$

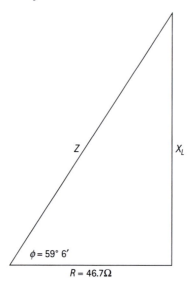

21 Find W and VAr in Figure 34.

Figure 34 Find W and VAr

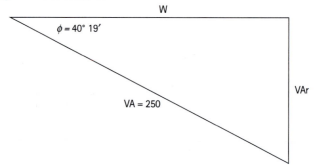

22 Find φ and X_L in Figure 35.

Figure 35 Find φ and X_L

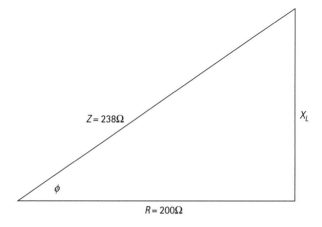

$Z = 238\Omega$

X_L

φ

$R = 200\Omega$

23 Find φ in Figure 36.

Figure 36 Find φ

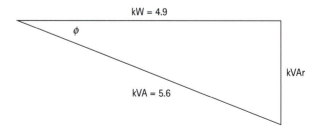

kW = 4.9

φ

kVAr

kVA = 5.6

24 Calculate *R* in Figure 37.

Figure 37 Find R

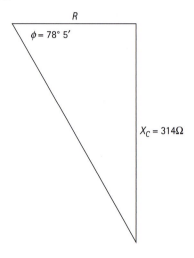

$\phi = 78° 5'$

$X_C = 314\Omega$

25 Find OX in Figure 38.

Figure 38 Find OX

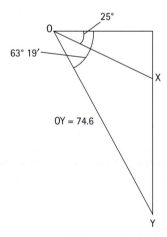

25°

63° 19'

OY = 74.6

26 Find OX in Figure 39.

Figure 39 Find OX

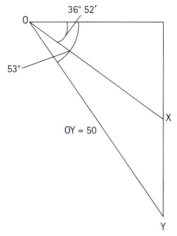

27 Complete the following table then plot the graph of p.f. (cos φ) to a base of phase angle (φ):

Phase angle φ			65°6′	60°			45°40′	
Power factor cos φ	0.25	0.3			0.55	0.6		0.82

28 A coil has inductance 0.18H and resistance 35Ω. It is connected to a 100V 50Hz supply. Calculate (a) the impedance of the coil, (b) the current which flows, (c) the power factor, (d) the power absorbed by the coil.

29 Define the term 'power factor' and state how it affects cable size. An inductor of resistance 8Ω and of inductance 0.015H is connected to an a.c. supply at 230V, single-phase, 50Hz. Calculate (a) the current from the supply, (b) the power in the circuit, (c) the power factor.

30 A single-phase a.c. supply at 230V 50Hz is applied to a series circuit consisting of an inductive coil of negligible resistance and a non-inductive resistance coil of 15Ω. When a voltmeter is applied to the ends of each coil in turn, the potential differences are found to be 127.5V across the inductive coil, 203V across the resistance. Calculate (a) the impedance of the circuit, (b) the inductance of the coil, (c) the current in the circuit, and (d) the power factor.

31 On what factors do the resistance, reactance and impedance of an a.c. circuit depend, and how are these quantities related? The current in a single-phase circuit lags behind the voltage by 60°. The power in the circuit is 3600W and the voltage is 240V. Calculate the value in ohms of the resistance, the reactance and the impedance.

Waveform and Phasor Representation of Alternating Currents and Voltages

6

ALTERNATING E.M.F. AND CURRENT

The value and direction of the e.m.f. induced in a conductor rotating at constant speed in a uniform magnetic field (Figure 40(a)) vary according to the position of the conductor.

The e.m.f. can be represented by the displacement QP of the point P above the axis XOX, Figure 40(b). OP is a line which is rotating about the point O at the same speed as the conductor is rotating in the magnetic field. The length of OP represents the maximum value of the induced voltage. OP is called a *phasor*.

A graph, Figure 40(c), of the displacement of the point P plotted against the angle θ (the angle through which the conductor has moved from the position

Figure 40 Alternating e.m.f. and current

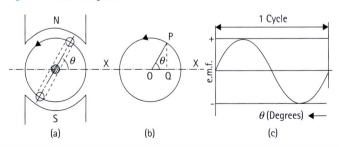

of zero induced e.m.f.) is called a *sine* wave, since the PQ is proportional to the sine angle θ. One complete revolution of OP is called a *cycle*.

EXAMPLE 1

An alternating voltage has a maximum value of 200V. Assuming that it is sinusoidal in nature (ie it varies according to a sine wave), plot a graph to show the variations in this voltage over a complete cycle.

Method (Figure 41) Choose a reasonable scale for OP; for instance, 10mm = 100V. Draw a circle of radius 20mm at the left-hand side of a piece of graph paper to represent the rotation of OP.

Figure 41 The rotation of OP

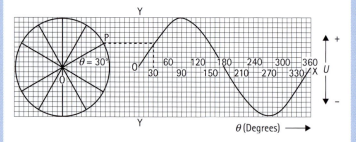

One complete revolution of OP sweeps out 360°. Divide the circle into any number of equal portions, say 12. Each portion will then cover 30°.

Construct the axes of the graph, drawing the horizontal axis OX (the x-axis) on a line through the centre of the circle. This x-axis should now be marked off in steps of 30° up to 360°. If desired, perpendicular lines can be drawn through these points. Such lines are called *ordinates*.

The points on the graph are obtained by projecting from the various positions of P to the coordinate corresponding to the angle θ at that position. Remember that when θ = 0° and 180° the generated e.m.f. is zero, and when θ = 90° and 270° the generated e.m.f. has its maximum value.

EXAMPLE 2

Two alternating voltages act in a circuit. One (A) has an r.m.s. value of 90V and the other (B) has an r.m.s. value of 40V, and A leads B by 80°. Assuming that both voltages are sinusoidal, plot graphs to show their variations over a complete cycle. By adding their instantaneous values together, derive a graph of the resultant voltage. Give the r.m.s. value of this resultant.

First find the maximum values of the voltages given:

$U_{r.m.s} = 0.707 \times U_{max}$

$\therefore 90 = 0.707 \times U_{max}$

$\therefore U_{max} = \dfrac{90}{0.707}$

$\qquad = 127\ V$

Similarly, if

$U_{r.m.s.} = 40$

$U_{max} = \dfrac{40}{0.707}$

$\qquad = 56.6\ V$

Choose a suitable scale, say 20mm = 100V. Draw two circles with the same centre, one having a radius of 25.4mm (127V), the other a radius of 11.32mm (56.6V).

Draw phasors to represent the voltages: OA horizontal and OB, which represents the lower voltage, lagging 80° behind OA (anticlockwise rotation is always used), see Figure 42.

Mark off the circumference of the larger circle in steps of 30°, using OA as the reference line.

Mark off the smaller circle in steps of 30°, using OB as the reference line.

Set off the axes of the graph alongside as in the previous example.

Plot the sine wave of voltage A as before.

Plot the sine wave of voltage B in exactly the same way, projecting the first point from B to the y-axis YOY and from each succeeding 30° point to the appropriate 30° point on the horizontal axis of the graph.

Figure 42 Phasors to represent the voltages

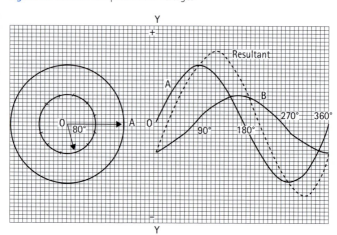

Points on the resultant graph are plotted by combining the ordinates of A and B at each 30° point. If the graphs lie on the same side of the x-axis, the ordinates are added. If the graphs lie on opposite sides of the axis, the smaller is subtracted from the larger (measurements upwards from the x-axis are positive, measurements downwards are negative).

EXAMPLE 3

A current of 15A flows from the 230V mains at a p.f. of 0.76 lagging.

Assuming that both current and voltage are sinusoidal, plot graphs to represent them over one cycle. Plot also on the same axes a graph showing the variation in power supplied over one cycle. The procedure for plotting the current and voltage sine waves is the same as that adopted in the previous example. The phase angle between current and voltage is found from the p.f. as follows:

p.f. = cosϕ

where ϕ is the angle of phase difference

cosϕ = 0.76

∴ ϕ = 40°32'

$$U_{max} = \frac{230}{0.707}$$

$$= 325.3V$$

$$I_{max} = \frac{15}{0.707}$$

$$= 21.21A$$

Scales of 20mm = 200V and 20 mm = 20A will be suitable. To obtain the graph of the power supplied, the ordinates of current and voltage are multiplied together (Figure 43). It is convenient to do this every 30° as before. Remember the rules for multiplying positive and negative numbers. Where the resulting graph is negative, additional points are helpful in obtaining a smooth curve.

Figure 43 Graph of the power

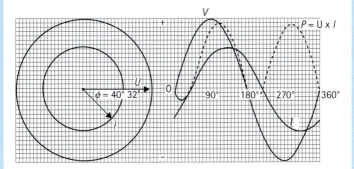

That portion of the power curve lying above the x-axis represents the power supplied to the circuit. That portion lying below the x-axis represents the power returned to the mains from the circuit.

EXERCISE 6

1 Plot a sine wave, over one complete cycle, of an alternating voltage having a maximum value of 325V. Determine the r.m.s. value of this voltage.

2 An alternating current has the following value taken at intervals of 30° over one half cycle:

Angle φ	0	30°	60°	90°	120°	150°	180°
Current A	0	10.5	17.5	19.7	15.0	11.5	0

Determine the average and r.m.s values of this current.

3 Two sinusoidal voltages act in a circuit. Their r.m.s. values are 110V and 80V and they are out of phase by 75°, the lower voltage lagging. Plot sine waves on the same axes to represent these voltages. Plot a graph of the resultant voltage by adding together the ordinates of the two waves. Give the r.m.s. value of the resultant voltage and state approximately the phase angle between this resultant and the lower voltage.

4 Two alternating currents are led into the same conductor. They are sinusoidal and have r.m.s. values of 4A and 1A. The smaller current leads by 120°. Plot out the sine waves of these two currents and add the ordinates to obtain the sine wave of the resultant current. Calculate the r.m.s. value of the resultant.

PHASORS

Conditions in alternating-current circuits can be represented by means of phasor diagrams.

In Figure 44, U is a voltage and I is a current, φ is the angle of phase difference, and cosφ is the p.f.

Figure 44 Phasor diagram

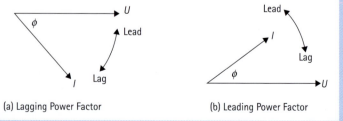

(a) Lagging Power Factor   (b) Leading Power Factor

EXAMPLE 1

The current in a circuit is 5A, the supply voltage is 230V and the p.f. is 0.8 lagging. Represent these conditions by means of a phasor diagram drawn to scale.

Figure 45 Choosing a scale

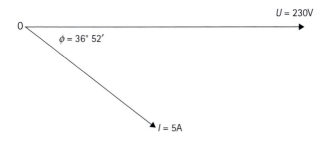

$U = 230V$

$\phi = 36°\ 52'$

$I = 5A$

Choose a suitable scale (see Figure 45).

p.f. = 0.8

   = cosφ

cosφ = 0.8

   φ = 36°52'

Normally the r.m.s. values are used when drawing phasor diagrams.

Note that the most accurate construction is obtained by setting off two lines at the required angle and then marking the lines to the appropriate lengths from the point of intersection with compasses which have been set to the respective measurement.

EXAMPLE 2

A resistor and a capacitor are wired in series to an a.c. supply (Figure 46). When a voltmeter is connected across the resistor it reads 150V. When it is connected to the capacitor terminals it indicates 200V. Draw the phasor diagram for this circuit to scale and thus determine the supply voltage. As the value of current is not given, it will not be possible to draw its phasor to scale. The current is the same throughout a series circuit and so the current phasor is used as a reference.

Figure 46 A resistor and a capacitor

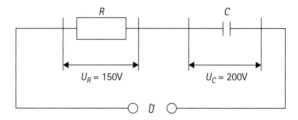

Draw OI any length to represent the current (Figure 47).

From point O, draw thin lines parallel to and at right angles to OI (capacitor voltage *lags* behind the current).

Choose a suitable scale and use compasses set to the required measurement to mark off OA = U_R, the resistor voltage drop (in phase with the current) and OB = U_C, the capacitor voltage drop.

With centre A and compasses set to length OB, strike an arc. With centre B and compasses set to OA, strike another arc. These arcs intersect at point C.

Figure 47 Draw OI

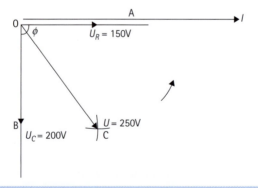

OC is the resultant voltage, which is equal to the supply voltage.

By measurement of OC, the supply voltage is found to be 250V.

EXAMPLE 3

An inductor takes a current of 5A from a 230V supply at a p.f. of 0.4 lagging. Construct the phasor diagram accurately to scale and estimate from the diagram the resistance and reactance of the coil. As already explained, although resistance and reactance cannot be separated, it is convenient to draw them apart in an equivalent circuit diagram (Figure 48). The total voltage drop (in this case the supply voltage) will then be seen to be made up of a resistance voltage drop and a reactance voltage drop.

Since, again, we are considering a series circuit in which the current is the same throughout, it is not necessary to draw the current phasor to scale.

Power factor = cosϕ where ϕ is the angle of phase difference between current and supply voltage and cosϕ = 0.4

∴ ϕ = 66°25′

Draw OI any length to represent the current (Figure 49).

Figure 48 Drawing resistance and reactance

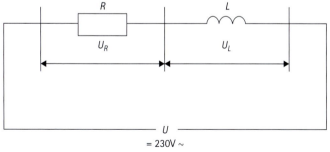

Equivalent Circuit Diagram

Figure 49 Parallel circuit

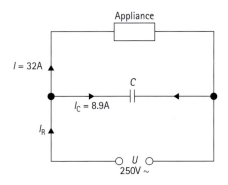

Choose a suitable scale and set off OC at 66°25' from OI and of length to represent the supply voltage.

Draw OY at right angles to the current phasor and from C draw perpendiculars to cut the current phasor at A and OY at B. The perpendiculars are constructed as follows:

(i) Set the compasses to any radius and with centre C draw arcs which cut OY at P and Q.

(ii) With the compasses again set to any radius and with centres P and Q strike two more arcs to cut in R. CR is then perpendicular to OY.

A similar method is employed in drawing CA.

By measurement,

$U_R = 93V$

$U_L = 209V$

Now $U_R = I \times R$

$93 = 5 \times R$

$\therefore R = \dfrac{93}{5}$

$= 18.6\Omega$

and $U_L = I \times X_L$ (X_L is the inductive reactance)

$\therefore\ 209 = 5 \times X_L$

$\therefore X_L = \dfrac{209}{5}$

$\qquad = 41.8\Omega$

EXAMPLE 4

An appliance takes a single-phase current of 32A at 0.6 p.f. lagging from a 250V a.c. supply. A capacitor which takes 8.9A is wired in parallel with this appliance (Figure 49). Determine graphically the total supply current. As this is a parallel circuit, the voltage is common to both branches and is thus used as the reference phasor. It need not be drawn to scale. Choose a suitable scale.

p.f. $= \cos\phi = 0.6$

$\therefore\ \phi = 53°8'$

Draw the voltage phasor (Figure 50) and set off the appliance-current phasor at 53°8' lagging (OA). The capacitor current, 8.9A, leads on the voltage

Figure 50 Voltage phasor

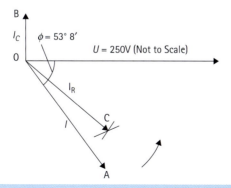

by 90° and is drawn next (OB). The resultant of these two phasors is found as follows:

(i) With compasses set to OA and centre B, strike an arc.
(ii) With centre A and compasses set to OB, strike another arc cutting the first in C.

OC is the resultant current. By measurement of OC, the resultant current is 25.5A.

EXAMPLE 5

A consumer's load is 15kVA single-phase a.c. at 0.8 p.f. lagging. By graphical construction, estimate the active and reactive components of this load.

p.f. = $\cos\phi$ = 0.8

$\therefore \phi = 36°52'$

Choose a suitable scale. Draw a thin horizontal line OX (Figure 51). Set off OA to represent 15kVA at an angle of 36°52' from OX.

From A, draw a perpendicular to cut line OX at B. OB is then the working or active component and AB is the reactive or wattless component.

Figure 51 Draw OX

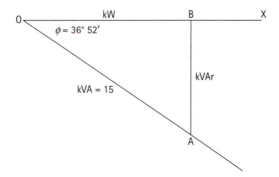

By measurement of OB the true power is 12kW, and by measurement of AB the wattless component is 9kVAr.

EXERCISE 7

1   A capacitor is wired in series with a resistor to an a.c. supply. When a voltmeter is connected to the capacitor terminals it indicates 180V. When it is connected across the resistor it reads 170V. Construct the phasor diagram for this circuit accurately to scale and from it determine the supply voltage.

2   A contactor coil takes a current of 0.085A from a 250V supply at a p.f. of 0.35 lagging. Draw the phasor diagram accurately to scale and use it to determine the resistance and reactance of the coil.

3   A single-phase transformer supplies 10kVA at 0.7 p.f. lagging. Determine by graphical construction the active and reactive components of this load.

4   The true power input to a single-phase motor is 1150W and the p.f. is 0.54 lagging. Determine graphically the apparent power input to the machine.

5   A fluorescent-lamp circuit takes a current of 1.2A at 0.65 p.f. lagging from the 230V a.c. mains. Determine graphically the true power input to the circuit.

6   A single-phase motor takes 8.5A from a 230V supply at 0.4 p.f. lagging. A capacitor which takes 4A is connected in parallel with the motor. From a phasor diagram drawn accurately to scale, determine the resultant supply current.

7   A discharge lighting fitting takes a current of 5.2A at 0.46 p.f. lagging when it is used without its power-factor-improvement capacitor. When this capacitor is connected the current falls to 3.2A, the supply voltage remaining constant at 240V. Draw the phasor diagram to represent the conditions with and without the capacitor and from it determine the current taken by the capacitor. (Remember that the working component of the supply current is constant.)

8   A series circuit is made up of a resistor, an inductor of negligible resistance, and a capacitor. The circuit is connected to a source of alternating current, and a voltmeter connected to the terminals of each component in turn indicates 180V, 225V and 146V, respectively. Construct the phasor diagram for this circuit accurately to scale and hence determine the supply voltage.

Parallel Circuits Involving Resistance, Inductance and Capacitance

7

Circuit with inductance and capacitance in parallel (Figure 52).

L is pure inductance (*henry*).

C is pure capacitance *(microfarad).*

In a parallel circuit the voltage is common to each branch of the circuit.

The current through the inductive branch is $I_L = \dfrac{U}{X_L}$ where $X_L = 2\pi f L$. This current lags the voltage by 90° (Figure 53a).

The current through the capacitive branch is $I_C = \dfrac{U}{X_C}$ where

$X_C = \dfrac{10^6}{2\pi f C}$ the current leads the voltage by 90° (Figure 53b).
Voltage is the reference and a current phasor is needed.

Figure 52 Inductance and capacitance in parallel

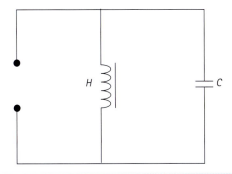

Figure 53a Lagging

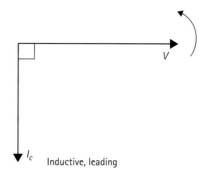

Inductive, leading

Figure 53b Current leads the voltage by 90°

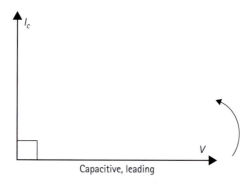

Capacitive, leading

EXAMPLE 1

Calculate the current drawn from the supply when an inductor with a reactance of 83Ω and a capacitor of 125Ω are connected in parallel to a 100V supply.

Capacitor current $I_c = \dfrac{U}{X_c}$

$$= \dfrac{100}{125}$$

$$=0.8A$$

Inductor current $\quad I_L = \dfrac{U}{X_L}$

$$= \dfrac{100}{83}$$

$$= 1.2A$$

Because inductor current is larger overall, the circuit is a lagging one.

The supply current is calculated $I_L - I_C$

$$= 1.2 - 0.8$$

$$= 0.4A$$

Lagging the voltage by 90°.

EXAMPLE 2

Calculate the current drawn from the supply when a capacitor of 75μF is connected in parallel with a resistor of 70Ω to a 110V 50Hz supply.

Draw a phasor diagram and determine the phase relationship between the supply voltage and the current drawn from the supply (Figure 54).

Figure 54 Phase relationship

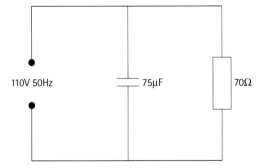

$$X_C = \frac{10^6}{2\pi f C}$$

$$= \frac{10^6}{2\pi \times 50 \times 75}$$

$$= 42.44\Omega$$

Enter into calculator EXP or $\times 10^x$ 6 ÷ (2 $shift$ π × 50 × 75) = (Ans)

$$I_C = \frac{110}{42.44}$$

$$= 2.59A$$

$$I_R = \frac{110}{70}$$

$$= 1.57A \text{ (see Figure 55).}$$

Figure 55 Graph of Example 2

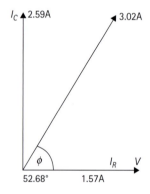

Find supply current by calculation $I_S^2 = I_C^2 + I_R^2$

$$= I_S = \sqrt{I_C + I_R}$$

$$= \sqrt{2.59 + 1.57}$$

$$= 3.02A$$

Enter into calculator 2.59 X^2 + 1.57x^2 = $\sqrt{}$ = (Ans)

To find phase angle by calculation

$$\phi = \frac{I_R}{I_S}$$

$$= \frac{1.57}{3.02}$$

$$= 0.52$$

$\phi 58.7°$

Enter into calculator $1.57 \div 3.02 = shift\ cos^- = (Ans)$

The current is leading the supply voltage by 58.7°.

EXAMPLE 3

A coil has a resistance of 25Ω and an inductive reactance of 20Ω. It is connected in parallel with a capacitor of 40Ω reactance to a 230V supply. Calculate the supply current and the overall p.f.

The coil impedance Z_L is

$$Z_L = \sqrt{R^2 + X_L^2}$$

$$= \sqrt{25^2 + 20X^2}$$

$$= 32.02\Omega$$

Coil current $I_L = \frac{U}{Z_L}$

$$= \frac{230}{32.02}$$

$$= 7.183A$$

Capacitor current $I_c = \frac{U}{X_c}$

$$= \frac{230}{40}$$

$$= 5.75A$$

Phase angle may be calculated $\cos \phi = \frac{R}{Z_L}$

$$= \frac{25}{32.02}$$

$$= 0.78$$

Cosø = 0.78

This is the p.f. of the coil alone and is lagging.

To find phase angle enter into calculator: shift cos⁻ 0.78 = (ans 38.7°)

Horizontal component of the coil current is $I_L \times cos$ø

= 7.2 × 0.78 = 5.61

Vertical component of coil = $\sqrt{7.2^2 - 5.61^2}$

$$= 4.51A$$

Enter on calculator $7.2^2 - 5.6^2 = \sqrt{}$ = (*Ans*)

Vertical component of capacitor current = 5.75A

Total vertical current = capacitor current − coil current

= 5.75 − 4.51

= 1.24A

$I^S = \sqrt{5.62^2 + 1.24^2}$

= 5.75A (see Figure 56).

Figure 56 Diagram of Example 3

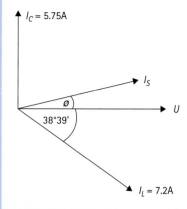

EXERCISE 8

1 Determine the current *I* in Figure 57 and state whether it leads or lags the voltage *U*.

Figure 57 Does it lag or lead?

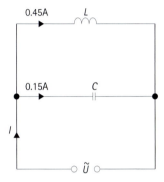

2 Determine the resultant current *I* and its phase relationship with the supply voltage *U* in Figure 58. What is the p.f. of the circuit?

Figure 58 What is the power factor?

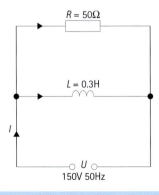

3 A capacitor of 15µF is connected in parallel with a coil of inductance 0.3H and negligible resistance to a sinusoidal supply of 240V 50Hz. Calculate the resultant current and state whether the phase angle is a leading or lagging one.

4 Calculate the resulting supply current and the overall p.f. when a resistor of 100Ω is connected in parallel with the circuit of Question 3.

5 A coil of reactance 30Ω and resistance 40Ω is connected in parallel with a capacitor of reactance 200Ω, and the circuit is supplied at 200V. Calculate the resultant current and p.f. Check the results by constructing the phasor diagram accurately to scale.

6 A coil has resistance 150Ω and inductance 0.478H. Calculate the value of a capacitor which when connected in parallel with this coil to a 50Hz supply will cause the resultant supply current to be in phase with the voltage.

7 An inductor coil of resistance 50Ω takes a current of 1A when connected in series with a capacitor of 31.8µF to a 240V 50Hz supply. Calculate the resultant supply current when the capacitor is connected in parallel with the coil to the same supply.

8 The resultant current in Figure 59 is

<blockquote>
(a) 0.585A   (b) 0.085A   (c) 11.2A   (d) 171A
</blockquote>

Figure 59 What is the resultant current?

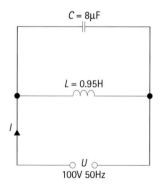

$C = 8µF$

$L = 0.95H$

I

U

100V 50Hz

9 The resultant current in Figure 60 is

    (a) 4A   (b) 8.5A   (c) 2.92A   (d) 9.22A

Figure 60 What is the resultant current?

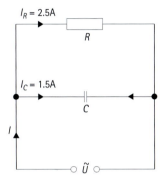

Power Factor Improvement

8

EXAMPLE 1

A consumer takes a load of 50kVA at 0.7 p.f. lagging. Calculate (a) the active and reactive components of the load, (b) the leading kVAr taken from a capacitor to improve the p.f. to 0.9 lagging.

(a) Active component (true power) $\dfrac{kW}{kVA} = pf$

Transpose to find kW

$kW \times 0.7 = 35 \ kW$

Reactive component $kW^2 = kVA^2 - kVAr^2$

Or $kVAr = \sqrt{kW^2 - kVA^2}$

$\qquad = \sqrt{50^2 - 35^2}$

$\qquad = 35.7 \ kVAr$

(b) Leading kVAr required

$\dfrac{kW}{kVA} = pf$

Transpose for kVA

$\dfrac{kW}{0.9} = 38.8$

$\dfrac{35}{0.9} = 38.8$

$kVAr = \sqrt{38.88^2 - 35^2}$

$\qquad = 16.93kVAr$

Figure 61 Power factor improvement

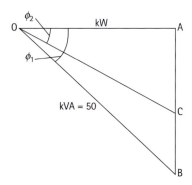

Lagging kVAr – leading kVAr = kVAr taken by capacitor

= 35.7 – 16.93 = 18.77 kVAr (see Figure 61).

EXAMPLE 2

A test on an 80W fluorescent lamp circuit gave the following results when connected to a 50Hz mains supply (Figure 62).

Figure 62 Diagram for Example 2

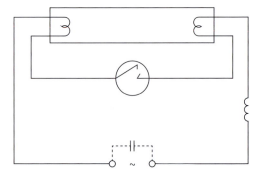

Without p.f. improvement capacitor.

Volts 232

Amperes 1.13

Watts 122

With p.f. correction capacitor

Volts 232

Amperes 0.68

Watts 122

Calculate the value of the p.f. correction capacitor in microfarads (μF).

The in phase current of the circuit is calculated $I = \dfrac{P}{U}$

$$= \dfrac{122}{232}$$

$$= 0.525A$$

This current is common to both cases since watts are the same (Figure 63).

Figure 63 Current stays the same as watts are the same

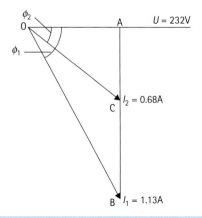

CALCULATION WITHOUT P.F. CORRECTION

Current drawn from supply = 1.13A

Wattless current in the reactive component:

$$= \sqrt{1.13^2 - 0.525^2}$$

$$= 1A$$

Enter into calculator $1.132x^2 - 0.525x^2 = \sqrt{} =$ (Ans)

CALCULATION WITH P.F. CORRECTION

Current drawn from supply = 0.68A.

Wattless current in reactive component:

$$= \sqrt{0.68^2 - 0.52^2}$$

$$= 0.432A$$

Difference between wattless current in circuit without capacitor and circuit with p.f. correction capacitor:

$$= 1A - 0.432A$$

$$= 0.568A$$

Calculate reactance of capacitor

$$U = I \times X_c$$

$$232 = 0.568 \times X_c$$

Transpose for X_c

$$\frac{232}{0.568} = 408\Omega$$

$$X_c = 408\Omega$$

For capacitance in microfarads

$$X_c = \frac{10^6}{2\pi f C}$$

Transpose for C

$$C = \frac{10^6}{2\pi f X_L}$$

$$= \frac{10^6}{2 \times 3.142 \times 50 \times 408}$$

Enter into calculator EXP or $\times 10^x$ 6 ÷ (2 $shift$ $\pi \times 50 \times 408$) = (Ans)

$= 7.8\ \mu F$

EXERCISE 9

1. The nameplate of a single-phase transformer gives its rating as 5kVA at 230V. What is the full-load current that this transformer can supply and what is its power output when the load p.f. is (a) 0.8, (b) 0.6?

2. (a) What is meant by power factor?

    (b) The installation in a factory carries the following loads: lighting 50kW, heating 30kW and power 44.760kW. Assuming that the lighting and heating loads are non-inductive, and the power has an overall efficiency of 87% at a p.f. of 0.7 lagging, calculate

      (i) the total loading in kW,

      (ii) the kVA demand at full load.

3. The current taken by a 230V 50Hz, single-phase induction motor running at full load is 39A at 0.75 p.f. lagging. Calculate the intake taken from the supply (a) in kW, (b) in kVA. Find what size capacitor connected across the motor terminals would cause the intake in kVA to be equal to the power in kW.

4. A group of single-phase motors takes 50A at 0.4 p.f. lagging from a 230V supply. Calculate the apparent power and the true power input to the motors. Determine also the leading kVAr to be taken by a capacitor in order to improve the p.f. to 0.8 lagging.

5. A welding set takes 60A from a 230V a.c. supply at 0.5 p.f. lagging. Calculate its input in (a) kVA, (b) kW. Determine the kVAr rating of a capacitor which will improve the p.f. to 0.9 lagging. What total current will now flow?

6. Explain with the aid of a phasor diagram the meaning of p.f. in the a.c. circuit. Why is a low p.f. undesirable? A single-phase load of 20kW at a p.f.

of 0.72 is supplied at 240V a.c. Calculate the decrease in current if the p.f. is changed to 0.95 with the same kW loading.

7. An induction motor takes 13A from the 240V single-phase 50Hz a.c. mains at 0.35 p.f. lagging. Determine the value of the capacitor in microfarads which, when connected in parallel with the motor, will improve the p.f. to 0.85 lagging. Find also the supply current at the new p.f.

8. A consumer's load is 100kVA at 0.6 p.f. lagging from a 240V 50Hz supply. Calculate the value of capacitance required to improve the p.f. as shown in the table below:

Power factor	0.7	0.75	0.8	0.85	0.9	0.95	1.0
Capacitance required (μF)							

9. An appliance takes a current of 45A at 0.2 p.f. lagging. Determine the current to be taken by a bank of capacitors in order to improve the p.f. to 0.6 lagging. Calculate the value of the capacitors in microfarads if they are supplied at (a) 240V, (b) 415V, and the supply frequency is 50Hz.

10. A test on a mercury vapour lamp gave the following results:

   Without power-factor-improvement capacitor: volts 230, amperes 2.22, watts 260

   With power-factor-improvement capacitor: volts 230, amperes 1.4, watts 260

   The supply frequency was 50Hz. Calculate the value of the capacitor in microfarads.

11. A transformer is rated at 10kVA 230V. The greatest current it can supply at 0.8 p.f. is

   (a) 43.3A

   (b) 34.8A

   (c) 23A

   (d) 230A

12. The power output of the transformer of question 11 at 0.8 p.f. is

    (a) 8kW

    (b) 12.5kW

    (c) 19.2kW

    (d) 3kW

13. A single-phase circuit supplies a load of 20kVA at 0.8 p.f. lagging. The kVAr rating of a capacitor to improve the p.f. to unity is

    (a) 16

    (b) 12

    (c) 25

    (d) 33.3

14. In order to improve the p.f., a circuit requires a capacitor to provide 6kVAr at 230V 50Hz. Its value in microfarads is

    (a) 1430µF

    (b) 143µF

    (c) 361µF

    (d) 3460µF

Three-phase Circuit Calculations

9

STAR-CONNECTED MOTORS

Three-phase supplies to an installation are normally in star formation with an earthed star point, the earthed star point provides a zero potential within the system to give a single-phase facility (as shown in Figure 64).

The colour code and sequence for phases is L1 Brown, L2 Black, L3 Grey.

On a standard installation the voltage between any two phases is 400V, this is called the line voltage U_L and between any phase and neutral the voltage will be 231V.

Figure 64 Star connection

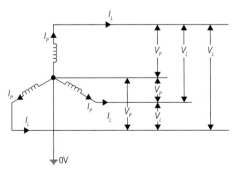

Calculation is $U_P = \dfrac{U_L}{\sqrt{3}}$

$$= \dfrac{400}{\sqrt{3}}$$

= 231V

The phasor for a balanced three-phase system is as shown in Figure 65.

The current in a three-phase star connected system is $I_L = I_P$ as shown in Figure 65.

I_L is the current in any line.
I_P is the current phase or load.

If the currents on a star-connected supply are the same on each phase, the system is said to be balanced. Under these circumstances the current in the neutral is zero.

The power per phase is P

$P = U^P \times I^P$

The total power is the sum of the power in each phase.
The total power in a balanced circuit can be calculated:

$P = \sqrt{3}\,U_L I_L$

Figure 65 A balanced three-phase system

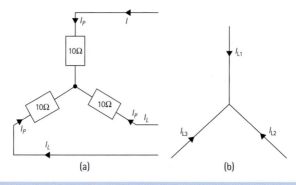

(a)   (b)

EXAMPLE

A three-phase balanced load is connected in star, each phase of the load has an impedance of 10Ω, the supply is 400V 50Hz.

Calculate the current in each phase (I_p):

$$I_p = \frac{U_p}{z \times \sqrt{3}}$$

$$= \frac{400}{10 \times 1.732}$$

$$= 23A \text{ per (phase } In \text{ } star \text{ } I_p = I_l)$$

Calculate the total power in one phase:

$$P = U_p \times I_p$$

$$= 231 \times 23$$

$$= 5313W$$

Total power in all three phases $5313 \times 3 = 15939W$.

Total power in all three phases can also be calculated:

$$P = U_L \times I_L \times \sqrt{3}$$

$$= 400 \times 23 \times 1.732$$

$$15.939W \text{ } (15.9kW)$$

DELTA CONNECTED MOTORS (MESH)

For delta connected loads (Figure 66) the voltage across the load will be the line voltage U_L, and the line current I_L will be the phase current I_p times $\sqrt{3}$.

Shown as a calculation $I_L = I_p \times \sqrt{3}$

The total power under these conditions is $P = \sqrt{3} \ U_L I_L$

EXAMPLE

(Using the same values as were used for star connection.)

Figure 66 Delta connected loads

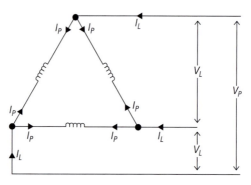

Figure 67 A balanced three phase load

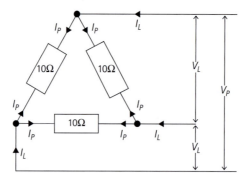

A three-phase balanced load is connected in delta, each phase of the load has an impedance of 10Ω and the supply is 400V 50Hz (Figure 67).

Calculate the phase current and the load current:

$$I_P = \frac{U_L}{Z}$$

$$= \frac{400}{10}$$

$$I_P = 40A$$

$I_L = I_P \times \sqrt{3}$

$= 40 \times 1.723$

$I_L = 69.28\ A$

The total power can be calculated:

$P = \sqrt{3} \times U_L \times I_L$

$= 1.732 \times 400 \times 69.28 = 47997\ Watts\ (47.8kw)$

It can be seen that the power dissipated in the delta-connected load is three times that of the star-connected load. The same applies to the current drawn from the supply.

RESISTANCE AND INDUCTANCE IN THREE–PHASE CIRCUITS

In many three-phase loads such as motors, inductance as well as resistance will need to be taken into account.

EXAMPLE

Three coils are connected in star formation to a 400V 50Hz supply, each coil has a resistance of 35Ω and an inductance of 0.07H.

Calculate (A) the line current I_L and (B) the total power dissipated.

Step 1
Calculate inductive reactance:

$X_L = 2\pi f L$

$= 2 \times 3.142 \times 50 \times 0.07$

$= 22Ω$

An impedance triangle could be drawn if required as follows:

Draw to scale a line representing resistance on the horizontal (opposite), at right angles to line R draw a line representing inductive reactance (adjacent). The length of the hypotenuse will represent the impedance Z (Figure 68).

87

Figure 68 Resistance and reactance

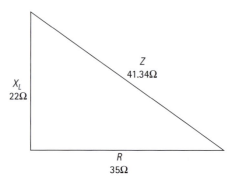

By calculation $Z^2 = X_L^2 + R^2$

$Z^2 = 22^2 + 35^2$

$Z = \sqrt{22^2 + 35^2}$

$\quad = 41.34\,\Omega$

Impedance (Z) is 41.34Ω per phase.

Because the circuit has inductive reactance and resistance, the load will have a p.f., this must now be calculated:

Power factor $= \dfrac{R}{Z}$

$\qquad = \dfrac{35}{41.34}$

$\qquad = 0.84$

Line current can now be calculated $I_L = \dfrac{U_L}{z \times \sqrt{3}}$

(U_L is line voltage)

$= \dfrac{400}{41.34 \times 1.732 \times 0.84}$

= 6.65A (In star $I_L = I_P$ remember this is the current per phase)

Power can now be calculated:

$$P = \sqrt{3} \times U_L \times I_L \times \cos\phi$$

$$= 1.732 \times 400 \times 6.65 \times 0.84$$

$$= 3867W \ (3.867 \ kW),$$

EXAMPLE 2

Using the coils in the previous example connected in delta, calculate the line current and total power dissipated.

$$\text{In delta} \quad I_p = \frac{U_L}{Z \times \cos\phi}$$

$$= \frac{400}{41.34}$$

$$I_p = 9.68A$$

$$I_L = I_p \times \sqrt{3}$$

$$= 9.68 \times 1.732$$

$I_L = 16.76A$(note: three times the current in as in star).

Power can now be calculated:

$$P = \sqrt{3} \times U_L \times I_L \times pf$$

$$= 1.732 \times 400 \times 16.76 \times 0.84$$

$$= 9750W \ (9.75kW) \ (note: three times the power as in star).$$

EXERCISE 10

1 Three equal coils of inductive reactance 30Ω and resistance 40Ω are connected in star to a three-phase supply with a line voltage of 400V. Calculate the line current and the total power.

2 The load connected between each line and neutral of a 400V 50Hz three-phase circuit consists of a capacitor of 31.8μF in series with a resistor of 100Ω. Calculate the current in each line and the total power.

3 The load connected between each line and the neutral of a 400V three-phase supply consists of:

between L1 and N, a non-inductive resistance of 25Ω;

between L2 and N, an inductive reactance of 12Ω in series with a resistance of 5Ω;

between L3 and N, a capacitive reactance of 17.3Ω in series with a resistance of 10Ω. Calculate the current in each line.

4 Three resistors each of 30Ω are connected in star to a 400V three-phase supply. Connected in star to the same supply are three capacitors each with a reactance of 40Ω. Calculate the resultant current in each line and the total power.

5 Three capacitors, each with a reactance of 10Ω are to be connected to a three-phase 400V supply for p.f. improvement. Calculate the current in each line if they are connected (a) in star, (b) in mesh.

6 A 440V, three-phase, four-wire system supplies a balanced load of 10kW. Three single-phase resistive loads are added between lines and neutrals as follows: L1 – N 2kW, L2 – N 4kW, L3 – N 3kW. Calculate the current in each line.

7 Three 30Ω resistors are connected in (a) star, (b) in delta to a 400V three-phase system. Calculate the current in each resistor, the line currents and the total power for each connection.

8 Each branch of a mesh connected load consists of a resistance of 20Ω in series with an inductive reactance of 30Ω. The line voltage is 400V. Calculate the line currents and total power.

9 Three coils, each with a resistance of 45Ω and an inductance of 0.2H, are connected to a 400V three-phase supply at 50Hz (a) in mesh, (b) in star. Calculate (i) the current in each coil and (ii) the total power in the circuit.

10 A three-phase load consists of three similar inductive coils, each with a resistance of 50Ω and an inductance of 0.3H. The supply is 400V 50Hz. Calculate (i) the current in each line, (ii) the p.f., (iii) the total power, when the load is (a) star-connected, (b) delta-connected.

11 Three equal resistors are required to absorb a total of 24kW from a 400V three-phase system. Calculate the value of each resistor when they are connected in (a) star, (b) mesh.

12 To improve the p.f., a certain installation requires a total of 48kVAr equally distributed over the three phases of a 415V 50Hz system. Calculate the values of the capacitors required (microfarads) when the capacitors are connected in (a) star, (b) delta.

13 The following loads are connected to a three-phase 400V 50Hz supply. A non-inductive resistance of 60Ω is connected between L1 and L2, an inductive reactance of 30Ω is connected between L2 and L3, and a capacitor of 100μF is connected between L1 and L3. Calculate the total power and the current through each load.

14 A motor generator set consists of a d.c. generator driven by a three-phase a.c. motor. The generator is 65% efficient and delivers 18A at 220V. The motor is 75% efficient and operates at 0.5 p.f. lagging from a 415V supply. Calculate (a) the power output of the driving motor, (b) the line current taken by the motor.

15 A conveyor raises 1600kg of goods through a vertical distance of 5m in 20s. It is driven by a gear which is 55% efficient. Calculate the power output of the motor required for this work.

If a three-phase 400V motor with an efficiency of 78% is fitted, calculate the line current assuming a p.f. of 0.7.

16 A 400V three-phase star-connected alternator supplies a delta-connected induction motor of full load efficiency 87% and a p.f. of 0.8. The motor delivers 14 920W. Calculate (a) the current in each motor winding, (b) the current in each alternator winding, (c) the power developed by the engine driving the alternator, assuming that the alternator is 82% efficient.

17 A three-phase transformer supplies a block of flats at 230V line to neutral. The load is balanced and totals 285kW at 0.95 p.f. The turns ratio of the transformer primary to secondary is 44:1 and the primary side of the transformer is connected in mesh. Calculate (a) the primary line voltage, (b) draw a diagram and mark the values of the phase and line currents in both windings.

THREE-PHASE CIRCUITS

In a balanced three-phase circuit no current will flow in the neutral.

In an unbalanced three-phase circuit some current will flow in the neutral, this current can be calculated by using four different methods.

EXAMPLE 1

A sub-main supplying an unbalanced three-phase and neutral distribution board has currents of 75A in L1 (brown), 55A in L2 (black) and 40A in L3 (grey). Calculate the current in the neutral (Figure 69).

Draw L1, L2, L3 to scale at 120 degrees to each other.

Now draw a vertical line from the end of L2 the same length as L1. Then join the tops of L1 and L2.

Draw a line between the shortest angles of parallelogram. Now draw a line the same length and parallel with L3 from top left angle. Join the ends of L3 and the line parallel to it.

Measure the gap between the shortest angle and if the phasor is drawn to scale this will be the current flowing in the neutral.

Figure 69 Calculate the current

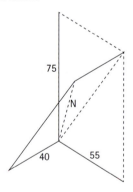

EXAMPLE 2

Using a triangle (Figure 70), draw a horizontal line to scale to represent L1.

From the right-hand side of this line, draw a line at 120 degrees to it to scale to represent L2. Now draw a line from the end of L2 at 120 degrees to it to scale to represent L3.

Figure 70 Diagram for Example 2

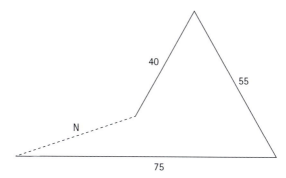

Measure the gap between the open ends of L1 and L3, this will be the current flowing in the neutral.

EXAMPLE 3

Using a simpler phasor diagram and a simple calculation (Figure 72).

Take the smallest current from the other two currents.

In this example L3 (40A) is the smallest current.

($L1$) 75A – 40A = 35A

($L2$) 55A – 40A = 15A

Figure 71 Phasor diagram

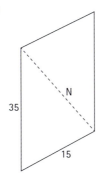

Draw a vertical line to scale to represent corrected $L1$.

Now draw a line to scale at 120 degrees to it to represent corrected $L2$.

Draw a line vertically from the end of $L2$ the same length as $L1$, join open ends.

Measure between the shortest angles and this will be the current in the neutral.

EXAMPLE 4 BY CALCULATION ONLY

Subtract the smallest current from the other two.

$(L1)75A - (L3) 40A - 35A$

$(L2)55A - (L3) 40A = 15A$

Current in the neutral can now be calculated:

$\sqrt{35^2 - 15^2} = 31.62$

EXERCISE 11

1 Three separate single-phase loads are to be connected to a three-phase and neutral supply, the currents in the loads are as follows: $L1 = 32A$, $L2 = 24A$ and $L3$ has 30A. Calculate the current flowing in the neutral.

2 A sub-main is to be installed to supply the following loads: $L1$ is a lighting load of 3.2kW, $L2$ is a cooker load of 7kW, $L3$ is supplying a 20A power load. The supply voltage is 230V 50H.

3 Calculate the current flowing in the neutral conductor of a supply cable when the following currents are flowing the phase conductors, L1 10A, L2 30A, L3 20A.

THREE-PHASE POWER

The chapter on power in book 1 explains leading and lagging p.f. In this chapter we will take p.f. a step further and see how it affects three-phase circuits.

To help it is often simpler to draw a power triangle showing:

Active power or true power in watts (*W*) *or kW*.

Apparent power in volt amps (*VA*) *or kVA*.

Reactive power in *VAr or kVAr*.

Power factor cosϕ is found $\dfrac{kW}{kVAr}$ (Figures 72 and 73).

Leading power factor

Figure 72 Leading

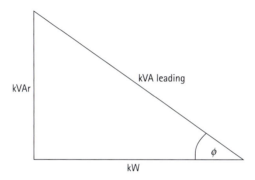

Lagging power factor

Figure 73 Lagging

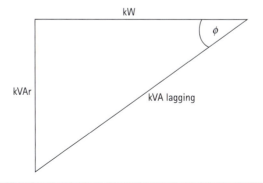

To find the true power, apparent power and the reactive power we must first calculate as follows:

EXAMPLE

A three-phase motor with an output of 2.8kW and a p.f. of 0.89 is connected to a 400V 50Hz supply. Calculate: (a) the power drawn from the supply, (b) the reactive power and (c) the line current.

A. $pf = \dfrac{kW}{kVA}$

$\quad$ = transposed $kVA = \dfrac{kW}{pf}$

$\quad = \dfrac{2.8}{0.89}$

= 3.14 kVA (power drawn from supply is greater the the power delivered by the motor)

B. $kVAr^2 = kW^2$ or $kVAr = \sqrt{kVA^2} - \sqrt{kW^2}$

$\quad = \sqrt{3.142^2 - 2.8^2}$

$\quad = 1.42$ kVAr

Enter into calculator $3.142X^2 - 2.8X^2 = \sqrt{\quad}$ = (Ans)

C. Line current calculation

$P = \sqrt{3}U_L I_L \cos\phi$

$I = \dfrac{W}{\sqrt{3} \times U_L \times \cos\phi}$

transposed

$\quad = \dfrac{2800}{\sqrt{3} \times 400 \times 0.89}$

$\quad = 4.54$A

EXAMPLE 1

A commercial building is supplied by a three-phase four wire 400V 50Hz supply and the phases are loaded as follows:

$L1$ is taking 35 *kW* at unity p.f.

$L2$ is taking 40 *kVA* at 0.8 lagging p.f.

$L3$ is taking 60 *kVA* at 0.7 leading p.f.

Calculate the p.f. for the system.

phase $L1$ is unity p.f. (1) $pf = \dfrac{kW\ true\ power}{kVA\ apparent\ power}$

transposed $kVA = \dfrac{kW}{pf}$

$$= \frac{35}{1}$$

$$= 35\,kVA$$

This phase has no reactive power.

The circuit is purely resistive and can be shown by phasor (see Figure 74).

Figure 74 A purely resistive circuit

Phase $L2$ has a p.f. of (lagging).

Using formulae $pf = \dfrac{kW}{kVA}$ transposed to find kW

$$= pf \times kVA$$

$$= 0.8 \times 40$$

$$= 32\,kW.$$

Reactive (*kVAr*) power can be calculated $kVar = \sqrt{kVA^2 - kW^2}$

$$= \sqrt{40^2 - 32^2}$$

$$= 24\,kVAr.$$

The circuit is inductive (lagging) and can be shown by phasor (see Figure 75).

97

Figure 75 An inductive circuit

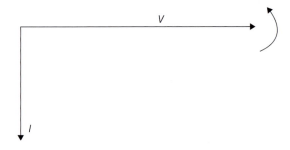

Phase $L3$ has a p.f. of leading.

Using the same formulae reactive power for phase $L3$ $pf \times kW$

$\qquad = 0.7 \times 60$

$\qquad = 42Kw$

$kVAr = \sqrt{60^2 - 42^2}$

$= 18kVAr$ leading.

This capacitive circuit can now be shown as a phasor (see Figure 76).

This circuit is capacitive and is a leading circuit.

Figure 76 A capacitive circuit

We must now decide if the system is lagging or leading. As the lagging component of the system is greater than the leading, it will be a lagging system.

The total kVAr will be the difference between the leading and lagging component.

Lagging 24$kVAr$
Leading 18$kVAr$
24 – 18 = 6$kVAr$.

Total power is calculated by adding the power $L1 + L2 + L3$

$L1 = 35kW$
$L2 = 32kW$ ($40A \times 0.8$)
$L3 = 42kW$ ($60A \times 0.7$)

Total power 109kW.

Total reactive component 6$kVAr$ (lagging).

kVA can now be calculated $\sqrt{6^2 + 109^2}$

= 109.16 kVA

p.f. can now be calculated $pf = \dfrac{kW}{kVA}$

$= \dfrac{109}{109.16}$

= 0.99

VOLTAGE DROP IN THREE–PHASE CIRCUITS

The method explained here is not a rigid treatment of three-phase voltage drop but will provide a result which will be sufficient for most purposes (*reactance is not taken into account*).

BS 7671 allows a maximum voltage drop of 3% for lighting circuits and 5% for all other circuits, this voltage drop is to be measured from the supply intake to the furthest point of the circuit.

In most installations the line voltage (UL) of a three-phase supply is 400V.

$$3\% \text{ of } 400 = \frac{400 \times 3}{100} = 12 \text{volts}$$

$$5\% \text{ of } 400 = \frac{400 \times 5}{100} = 20 \text{volts}$$

If the three-phase circuit is not a balanced load, each phase should be calculated as a single-phase circuit

$$3\% \text{ of } 230 = \frac{230 \times 3}{100} = 6.9 \text{volts}$$

$$5\% \text{ of } 230 = \frac{230 \times 5}{100} = 11.5$$

EXAMPLE 1

A three-phase balanced load of 15A per phase is supplied by a steel wire armoured cable with a c.s.a. of 2.5mm^2. The voltage drop for this cable is 15mv/A/m and the circuit is 40m long. Calculate (a) the voltage drop in the cable and (b) determine if it complies with the voltage drop requirements of BS 7671.

A. $Voltdrop = \dfrac{mV \times I \times L}{1000}$

$$= \frac{15 \times 15 \times 40}{1000}$$

= 9 volts per phase

The equivalent voltage drop in line voltage is

$$= \sqrt{3} \times 9$$

= 15.48 $volts$

B. This complies as it is less than 20 volts.

EXAMPLE 2

A three-phase 10kW motor operates on full load with efficiency 80% and p.f. 0.75. It is supplied from a switchboard through a cable each core of which has resistance 0.2Ω. Calculate the voltage necessary at the supply end in order that the voltage at the load end terminals shall be 400V.

The full load current of the motor is

$$I = \frac{10 \times 1000}{\sqrt{3} \times 400 \times 0.75} \times \frac{100}{80}$$

$$= 24.06A$$

The voltage drop per core of the cable is

24.06×0.2

$= 4.8V$

The equivalent reduction in the line voltage is

$\sqrt{3} \times 4.8$

$= 8.31V$

The required voltage at the switchboard is

$400 + 8.31$

$= 408.3V$

This is not a rigid treatment of the problem but the method gives a result sufficiently accurate for most practical purposes.

EXAMPLE 3

The estimated load in a factory extension is 50kW balanced at 0.8 p.f. The supply point is 120m away and the supply voltage is 400V. Calculate the cross-sectional area of the cable in order that the total voltage drop shall not exceed 2.5% of the supply voltage.

Take the resistivity of copper as 1.78×10^{-8} Ωm.

The line current

$$I = \frac{50 \times 1000}{\sqrt{3} \times 400 \times 0.8}$$

$$= 90.21A$$

Allowable reduction in line voltage

= 2.5% × 400

= 10V

Equivalent reduction in line voltage

$$= \frac{10}{\sqrt{3}}$$

= 5.77V

Resistance per core of the cable

$$= \frac{5.77}{90.21}$$

= 0.06396Ω

The resistance of a cable is given by

$$R = \frac{pl}{A}$$

where p is the resistivity (Ωm)

l is the length (m)

A is the cross-sectional area (m²) so that

$$A = \frac{pl}{R}$$

$$A = \frac{1.78\Omega\text{m} \times 120\text{m}}{10^8 \times 0.06396\Omega}$$

$$= \frac{0.33}{10^4}\text{m}^2$$

$$= \frac{0.33}{10^4}\text{m}^2 \frac{[16^6\,\text{mm}^2]}{\text{m}^2}$$

$$= 33\text{mm}^2$$

EXAMPLE 4

A length of pvc trunking is to be used to enclose single-core pvc-insulated distribution cables (copper conductors) for a distance of 30m from the main switchgear of an office building to supply a new 400V T.P. and N. distribution fuseboard. The balanced load consists of 24kW of discharge lighting. The fuses at the main switch-fuse and at the distribution board are to BS 88 Part 2. The voltage drop in the cables must not exceed 6V. The ambient temperature is anticipated to be 35°C. The declared value of Ip is 20kA and that of Z_e is 0.30Ω. Assume that the requirements of BS 7671 section 434 are satisfied by the use of BS 88 fuses.

(a) For the distribution cables, establish the:

  (i) design current (I_b)

  (ii) minimum rating of fuse in the main switch-fuse (I_n)

  (iii) maximum mV/A/m value

  (iv) minimum current rating (I_t)

  (v) minimum cross-sectional area of the live conductors

  (vi) actual voltage drop in the cables.

(b) It is proposed to install a 2.5mm² protective conductor within the pvc trunking. Verify that this meets shock protection requirements. (C & G)

  (a) (i) Design current $I_b = \dfrac{24 \times 10^3 \times 1.8}{\sqrt{3} \times 400}$ (1.8 factor for discharge lighting)

  = 62.36A

  (ii) Minimum BS 88 fuse rating (I_n) is 63A.

  (iii) Maximum mV/A/m value $= \dfrac{6 \times 1000}{62.36 \times 30}$

  = 3.2mV/A/m

  (iv) Minimum current rating $(I_t) = \dfrac{63}{0.94}$ (temperature correction factor C_a for 35°C)

  = 67.02A

  (v) Minimum c.s.a. of cable is 16mm² (68A 2.4 mV/A/m).

  (vi) Actual voltage drop in 30m $= \dfrac{2.4 \times 62.36 \times 30}{1000} = 4.49V$

  (b) Check compliance with Table 41.4 (BS 7671) using *IET On-Site Guide*.

From Table I1, $R_1 + R_2$ for $16mm^2/2.5mm^2 = 1.15 + 7.41 m\Omega/m$. From Table I3, factor of 1.20 must be applied.

Now $Z_s = Z_e + R_1 + R_2$

$$R_1 + R_2 = \frac{30 \times (1.15 + 7.41) \times 1.20}{1000}$$

$$= 0.308\Omega$$

$$\therefore Z_s = 0.3 + 0.308$$

$$Z_s = 0.608\Omega$$

This satisfies Table 41.4 as the maximum Z_s for a 63A fuse is 0.86Ω.

EXAMPLE 5

It is proposed to install a new 230V 50Hz distribution board in a factory kitchen some 40m distant from the supplier's intake position. It is to be supplied by two $25mm^2$ pvc insulated (copper conductors) single-core cables in steel conduit. Protection at origin of the cables is to be by BS 88 fuses rated at 80A.

It is necessary for contractual purposes to establish:

(a) the prospective short circuit current (p.s.c.c.) at the distribution board, and
(b) that the proposed distribution cables will comply with BS 7671 Requirement 434.

A test conducted at the intake position between phase and neutral to determine the external impedance of the supplier's system indicates a value of 0.12Ω.

(a) The resistance of distribution cables from intake to distribution board

From Table I1 (*IET On-Site Guide*), R_1/R_2 for $25mm^2/25mm^2$ cables = 1.454 mV/m.

From Table I3 a multiplier of 1.20 is necessary using the Table 6A figures as

$$R_1/R_2 = \frac{40 \times 1.454 \times 1.20}{1000}$$

$$= 0.0698\Omega \text{ (regard this as impedance)}$$

So total short circuit fault impedance = 0.12 + 0.0698

$$= 0.19\Omega$$

Thus $I_f = \dfrac{230}{0.19}$

∴ p.s.c.c. = 1210A

From Appendix 3, Figure 3.3A, the BS 88 fuse clearance time is approximately 0.1s.

(b) From Requirement 434.5.2, $t = \dfrac{k^2 S^2}{I^2}$

$$= \dfrac{115^2 \times 25^2}{1200^2}$$

∴ limiting time for conductors (t) = 5.74s.

The cables are disconnected well before the 25mm^2 cable conductors reach their limiting temperature, thus they are protected thermally.

EXAMPLE 6

Two 25mm^2 single-core pvc-insulated cables (copper conductors) are drawn into pvc conduit along with a 10mm^2 protective conductor to feed a 230V industrial heater. The following details are relevant:

Protection at the origin is by 80A BS 88 fuses.

The tested value of Z_e at the cables origin is 0.35Ω.

The length of cables run is 55m.

(a) Establish the:

    (i) value of $R_1 + R_2$ of the cables

    (ii) prospective earth fault loop current (I_{ef})

    (iii)disconnection time of the fuse.

(b) Does the clearance time comply with BS 7671?

    (a) (i) Using the *IET On-Site Guide*

From Table I1 $R_1 + R_2$ for 25mm²/10mm² cables = 2.557mΩ²/m. From Table I3 apply the factor 1.20

Thus $R_1 + R_2 = \dfrac{55 \times 2.557 \times 120}{1000}$

$\qquad = 0.169\Omega$

So Z_s at distribution board = 0.35 + 0.169

$\qquad\qquad\qquad\qquad = 0.519\Omega$

    (ii) Prospective earth fault current $(I_{ef}) = \dfrac{230}{0.519}$

    = 443A

    (iii) Using BS 7671, Appendix 3, Table 3.A3, disconnection time is
       3.8s.

    (b) The clearance time complies with BS 7671 Requirement 411.3.2
       which specifies a maximum disconnection time of 5s.

EXAMPLE 7

The declared value of I_p at the origin of a 230V 50Hz installation is 1.5kA. The length of 25mm² pvc/pvc metre tails is 2m; at this point a switch-fuse containing 100A BS 88 Part 2 fuses is to be installed to provide control and protection for a new installation. A 20m length of 16mm² heavy duty mineral insulated cable (exposed to touch) (copper conductors and sheath) is to be run from the switch-fuse to a new distribution board.

(a) Establish that the mineral cable complies with BS 7671 Requirement 434.5.

(b) How could you ensure that the requirements of BS 7671 411.3.2, etc and Chapter 7 are satisfied?

(a) Resistance of 2m of 25mm² metre tails using mV/A/m value from Table I1 as ohms per metre at 70°C

$R_{mt} = 2 \times 0.00175$

$\qquad = 0.0035\Omega$

Resistance of 20m of 16mm² twin m.i.c.c. cable using resistance values from Table I1 as ohms per metre at 20°C

$R_{mi} = 20 \times 0.0023$

$= 0.046\Omega \times 1.2 = 0.055\Omega$

Impedance of supply $= \dfrac{230}{1500}$

$= 0.153\Omega$

Thus total impedance from source to distribution board

$= 0.153 + 0.0035 + 0.055 = 0.211\Omega$

Prospective short circuit fault current

$I_p = \dfrac{230}{0.211} = 1090\text{A}$

Disconnection time from BS 7671 Table 3.B3 is 0.3s

Now using the 434.5.1 adiabatic equation

$t = \dfrac{k^2 S^2}{I^2}$

$= \dfrac{135^2 \times 16^2}{1090^2}$

$= 3.92\text{s}$

Thus 16mm^2 cable is protected against the thermal deterioration.

(b) As no details are available in BS 7671 in relation to the resistance/impedance of the m.i.c.c. sheath, the prospective value of Z_s could not be established, but the actual value must be tested when the installation is commissioned and the value recorded in the Electrical Installation Certificate referred to in Requirement 610.6.

EXAMPLE 8

A 230V, 50Hz, 5kW electric motor is fed from a distribution board containing BS 88 Part 2 fuses. The wiring between the d.f.b. and the motor starter which is 20m distant is pvc-insulated single-core cables drawn into steel conduit. Assume that the:

(i) starter affords overload protection;
(ii) motor has a p.f. of 0.75 and an efficiency of 80%;

(iii) ambient temperature is 40°C;

(iv) fuse in the d.f.b. may have a rating up to twice the rating of the circuit cables;

(v) voltage drop in the motor circuit cables must not exceed 6V;

(vi) resistance of metal conduit is 0.1Ω, per metre;

(vii) 'worst' conduit run is 8m with two 90° bends;

(viii) I_p at d.f.b. is 2kA;

(ix) value of Z_e is 0.19Ω.

Establish the:

(a) design current (I_b);

(b) rating of circuit fuse;

(c) minimum cable rating (I_n) between d.f.b. and starter;

(d) minimum cable cross-sectional area;

(e) actual voltage drop in cables;

(f) prospective short circuit current;

(g) short circuit disconnection time;

(h) whether BS 7671 Requirement 434.5, etc. is satisfied;

(i) whether BS 7671 Requirement 411.3.2, etc. is satisfied;

(j) minimum conduit size.

(a) Design current $(I_b) = \dfrac{5000}{230 \times 0.75 \times 0.8}$

$= 36.2A$

(b) Rating of circuit fuse may be 80A.

(c) Minimum cable rating may be 40A.

(d) Minimum cable c.s.a. $= \dfrac{40}{0.87} = 46A$ from Table 4D1A select 10mm^2 cable (57A).

(e) Actual voltage drop: from Table 4D1BmV/A/m value for 10mm^2 is 4.4 thus

voltage drop $= \dfrac{36.2 \times 20 \times 4.4}{1000}$

$= 3.19 \text{ V}$

(f) Impedance of supply cables to d.f.b.

$= \dfrac{230}{2000} = 0.115Ω$

Using BS 7671 Tables I1 and I3, resistance of circuit cables

$$= \frac{20 \times 3.66 \times 1.2}{1000} = 0.09\Omega$$

thus total circuit impedance = 0.115 + 0.09 = 0.205Ω

Prospective short circuit current $= \dfrac{230}{0.205} = 1122A$

(g) Disconnection time from Figure 3.3A is 0.1s

(h) Cable thermal capacity $t = \dfrac{k^2 \times S^2}{I^2} = \dfrac{115^2 \times 10^2}{1122^2}$

$$= 1.05s$$

Thus 10mm^2 cables are thermally safe.

(i) Now $Z_s = Z_e + R_1 + R_2$

Using BS 7671 Tables I1 and I3,

Resistance of R1 $= \dfrac{20 \times 1.83}{1000} = 0.0366\Omega$

Resistance of conduit $R_2 = 20 \times 0.01 = 0.2\Omega$

Thus $Z_s = 0.19 + 0.0366 + 0.2 = 0.4266\Omega$

$I_{ef} = \dfrac{230}{0.4266} = 539A$

From BS 7671, Figure 3.A3, disconnection time is 1.4s; this being less than 5s, protection is satisfactory,

(j) From Table E3, cable factor for 2 × 10mm^2 cables = 2 × 105 = 210

From Table E4 select 25mm conduit (factor 292).

EXERCISE 12

1 A balanced load of 30A is supplied through a cable each core of which has resistance 0.28Ω. The line voltage at the supply end is 400V. Calculate the voltage at the load end, the percentage total voltage drop and the power wasted in the cable.

2 Each core of a three-core cable, 164m long, has a cross-sectional area of 35mm^2. The cable supplies power to a 30kW, 400V, three-phase motor working at full load with 87% efficiency and p.f. 0.72 lagging. Calculate:

(a) the voltage required at the supply end of the cable; (b) the power loss in the cable.

The resistivity of copper may be taken as 1.78×10^8 Ωm and the reactance of the cable may be neglected.

3   A 40kW, 400V, three-phase motor, running at full load, has efficiency 86% and p.f. 0.75 lagging. The three-core cable connecting the motor to the switchboard is 110m long and its conductors are of copper 25mm² in cross-section.

Calculate the total voltage drop in the cable, neglecting reactance. If the cable runs underground for most of its length, choose a suitable type of cable for the purpose and give a descriptive sketch of the system of laying it. The resistivity of copper may be taken as 1.78×10^8 Ωm.

4   The estimated load in a factory extension is 200kW at 0.85 p.f. (balanced). The supply point is 75m away where the line voltage is 400V. Choose the most suitable size of cable from those given below in order that the total voltage drop shall not exceed 2.5% of supply voltage.
Cross-sectional areas of available conductors (mm²) 35   50   70   95
(Resistivity of conductor is 1.78×10^8 Ωm).

5   A motor taking 200kW at 0.76 p.f. is supplied at 400V three-phase by means of a three-core copper cable 200m long.

(a) Calculate the minimum cable cross-sectional area if the voltage drop is not to exceed 5V.

(b) If the cable size calculated is non-standard, select from the table a suitable standard cable and calculate the actual voltage drop using that cable.

Standard cross-sectional areas of cable conductors (mm²) 300   400   500   630
(Resistivity of copper 1.78×10^8 Ωm.)

6   A three-phase current of 35A is supplied to a point 75m away by a cable which produces a voltage drop of 2.2mV per ampere per metre. Calculate the total voltage drop.

The following question should be answered by reference to the appropriate tables in BS 7671 and/or in the *IET On-Site Guide* to BS 7671.

7   A balanced load of 85A is required at a point 250m distant from a 400V supply position. Choose a suitable cable (clipped direct) from Tables 4E4A

and 4E4B in order that the total voltage drop shall be within the BS 7671 specified limit (ambient temperature 30°C).

8 A 25kW, 400V three-phase motor having full load efficiency and p.f. 80% and 0.85 respectively is supplied from a point 160m away from the main switchboard. It is intended to employ a surface-run, multicore pvc-insulated cable, non-armoured (copper conductors). The ambient temperature is 30°C and BS 88 fuses are to be employed at the main switchboard. Select a cable to satisfy the BS 7671 requirements.

9 The total load on a factory sub-distribution board consists of: 10kW lighting balanced over three phases, unity p.f.; 50kW heating balanced over three phases, unity p.f. and 30kW motor load having an efficiency 80%, p.f. 0.8.

The line voltage is 400V and the supply point is 130m distant. Protection at the origin of the cable (clipped direct) is by BS 88 fuses. The ambient temperature is 30°C. Select a suitable cable from Tables 4D2A and 4D2B, in order that the voltage drop shall not exceed 3% of the supply voltage.

10 Calculate the additional load in amperes which could be supplied by the cable chosen for Question 9 with the voltage drop remaining within the specified limits.

11 A 12kW, 400V three-phase industrial heater is to be wired using single-core pvc-insulated cables (copper conductors) 30m in length drawn into a steel conduit. The following details may be relevant to your calculation.

Ambient temperature 40°C.

Protection by BS 3036 (semi-enclosed) fuses.

Voltage drop in the cables must not exceed 10V.

The contract document calls for a 2.5mm^2 conductor to be drawn into the conduit as a supplementary protective conductor.

The worst section of the conduit run involves two right-angle bends in 7m. Establish the:

    (a) design current (I_b);

    (b) minimum fuse rating (I_n);

    (c) maximum mV/A/m value;

(d) minimum live cable rating (I_t);

(e) minimum live cable c.s.a.;

(f) actual voltage drop;

(g) minimum conduit size.

12 The external live conductor impedance and external earth fault loop impedance are tested at the intake of a 230V single-phase installation and show values of 0.41Ω and 0.28Ω, respectively. A pvc trunking runs from the intake position to a distribution board 40m distant and contains 35mm² live conductors and a 10mm² protective conductor.

(a) Estimate the:

(i) prospective short circuit current (p.s.c.c.) at the distribution board;

(ii) p.s.c.c. clearance time of the 100A BS 88 fuse at the origin of the cable;

(iii) value of the earth fault loop impedance (Z_s) at the distribution board;

(iv) prospective earth fault loop current;

(v) earth fault clearance time of the BS 88 fuse at the origin of the cable.

(b) State the maximum permitted value of Z_s under these conditions.

13 A pvc trunking containing single-core pvc-insulated distribution cables (copper conductors) is to be run 30m from the 400/230V main switchgear of an office building to supply a new T.P. and N. distribution fuseboard. The balanced load consists of 24kW of discharge lighting. The fuses at the main switch-fuse and at the distribution board are to BS 88 Part 2. The voltage drop in the distribution cables must not exceed 6V and the ambient temperature is anticipated to be 35°C. The declared value of I_p is 20kA and that of Z_e is 0.3Ω. Assume that the requirements of BS 7671 434–5 are satisfied.

(a) For the distribution cables, establish and state the:

(i) design current;

(ii) minimum rating of fuse in the main switch fuse;

(iii) maximum mV/A/m value;

(iv) minimum current rating;

(v) minimum cross-sectional area of the live conductors;

(vi) actual voltage drop in the cable.

(b) It is proposed to instal a 4mm^2 protective conductor within the pvc trunking.

    (i) State the value of Z_s.

    (ii) Verify that this meets BS 7671 shock protection requirements.

14 A security building is to be built at the entrance to a factory. This new building is to be provided with a 230V single-phase supply and is to be situated 20m from the main switchroom. A 30m twin pvc-insulated armoured underground cable (copper conductors) supplies the new building, which allows 5m at each end for runs within the main switchroom and security buildings. The connected load in the security building comprises:

one 3kW convector heater

two 1kW radiators

two 1.5kW water heaters (instantaneous type)

one 6kW cooker

six 13A socket outlets (ring circuit)

a 2kW lighting load.

Diversity may be applied (business premises).

Establish:

    (a) the prospective maximum demand;

    (b) minimum current rating of the switch-fuse in the switchroom at the origin of the underground cable.

    (c) Determine

        (i)   the minimum current rating of pvc-insulated twin armoured (copper conductor) underground cable, assuming an ambient temperature of 25°C and protection to be by BS 88 Part 2 devices;

        (ii) the minimum size of cable, assuming the voltage drop is limited to 2V;

        (iii) the actual voltage drop in the cable.

15 It is proposed to instal a pvc-insulated armoured cable to feed a 25kW, 400V, three-phase 50Hz resistive element type of furnace. The cable is to be surface run along a brick wall of a factory and has a total length of 95m. The protection at the origin of the circuit is to be by BS 88 fuses. The cable armour may be relied upon as the circuit protective conductor. The ambient temperature in the factory will not exceed 35°C and the voltage drop must not exceed 10V.

Determine and state the:

(a) design current;

(b) fuse rating;

(c) minimum cable current rating;

(d) maximum mV/A/m value;

(e) minimum cross-sectional area of the live conductors;

(f) actual voltage drop in the cable.

16 A pipeline pump is connected to a 400/230V three-phase supply. It is wired in 1.5mm² pvc-insulated cables drawn into 25mm steel conduit running 30m from a distribution board containing BS 88 Part 2 fuses (10A fuses protect the pump).

It is now necessary to add alongside the pump a 12kW 400/230V heater and it is proposed to draw the pvc-insulated cables into the existing 25mm pump circuit conduit and insert suitable fuses into the distribution board which has vacant ways.

The following assumptions may be made:

(i) an ambient temperature of 35°C;

(ii) the maximum distance between draw in boxes is 9.5m with two right-angle bends;

(iii) the maximum voltage drop in the heater circuit is 5V;

(iv) a 2.5mm² protective conductor is installed in the steel conduit to satisfy a clause in the electrical specifications.

Determine for the heater circuit the:

(a) design current I_b;

(b) suitable fuse rating I_n;

(c) maximum mV/A/m value;

(d) minimum cable current rating I_t;

(e) minimum cable c.s.a.;

(f) actual voltage drop.

For the existing pump circuit establish whether the:

(g) pump circuit cable current rating is still adequate;

(h) 25mm conduit is suitable for the additional cables.

Voltmeters and Ammeters: Changing the Use and Extending the Range

10

VOLTMETERS

The voltmeter is a high-resistance instrument and its essential electrical features may be represented by the equivalent circuit of Figure 77.

Figure 77 A voltmeter

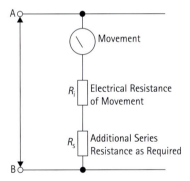

R_i is the 'internal resistance' of the movement, i.e. the resistance of the moving coil or the resistance of the fixed coil in the case of a moving-iron instrument. Independently of its resistance, the movement will require a certain current to deflect the pointer across the full extent of the scale against the effect of the controlling springs. This is the current required for full-scale deflection (f.s.d.).

The range of voltage which the instrument can indicate is governed by the total resistance R as measured between the terminals A and B, and

$R = R_i + R_s$

If $I_{f.s.d.}$ is the current required to produce full-scale deflection and R is the *total* resistance between A and B, the voltage between A and B at full-scale deflection is

$U = R \times I_{f.s.d.}$

$I_{f.s.d.}$ is fixed by the mechanical and electrical characteristics of the instrument and is not normally variable. The resistance R, however, can be fixed at any convenient value by adding the additional series resistance (R_s) as required.

EXAMPLE

An instrument has internal resistance 20Ω and gives f.s.d. with a current of 1mA. Calculate the additional series resistance required to give f.s.d. at a voltage of 100V.

$U = R \times I_{f.s.d.}$

$\therefore 100 = R \times \dfrac{1}{1000}$ (note the conversion of milliamperes to amperes)

$\therefore R = 100 \times 1000Ω$

$\quad = 100\ 000Ω$

But the instrument has internal resistance of 20Ω; thus the additional resistance required is

$R_s = R - R_i$

$\quad = 100\ 000 - 20$

$\quad = 99\ 980Ω$

(Note that this is a somewhat unrealistic value in terms of what it is economically practical to manufacture. In practice, the additional resistance would be constructed to the nominal value of 100 000Ω (100kΩ) and slight adjustments would be made as necessary at the calibration stage to obtain f.s.d. with an applied 100V.)

Any applied voltage less than 100V of course produces a corresponding lower reading on the instrument.

The additional series resistor R_s is also known as a *multiplier*.

AMMETERS

The ammeter is a low-resistance instrument, and its equivalent electrical circuit is shown in Figure 78.

Figure 78 An ammeter

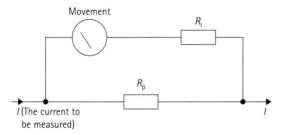

The resistor R_p connected in parallel is the 'shunt' through which most of the current to be measured flows. Its value will be low compared with the internal resistance of the movement R_i. The calculation of R_p proceeds as follows.

Knowing R_i and the current required to give f.s.d., determine the voltage required to produce f.s.d. of the movement. For example, using the information of the previous example ($I_{f.s.d.} = 1mA$ and $R_i = 20Ω$),

p.d. required for f.s.d. $= U_{f.s.d.} = I_{f.s.d.} \times R_i$

$$= \frac{1}{1000} \times 20$$

$$= \frac{20}{1000} V$$

and this is the voltage drop which must be produced across the shunt resistor R_p. The current which flows through R_p is the current to be measured minus the current which flows through the movement. If the greatest value of current to be measured is 20A, the current which flows through R_p is $I = 20A - 1mA = 19.999A$. The voltage across R_p is then

$$V_{f.s.d.} = (20/1000) \text{ V},$$

$$\therefore R_p = U_{f.s.d.} \over I$$

$$= \frac{20/1000}{19.999}$$

$$= 0.001\Omega$$

In fact, no significant difference is made if the 1mA of total current which flows through the movement and not through the shunt is ignored in the calculation of R_p. Again, it is usual to manufacture the shunt to the nominal value calculated above and then to make slight adjustments at the calibration stage to obtain the desired full-scale deflection.

EXAMPLE

A moving-coil instrument gives full-scale deflection with a current of 1.2mA, and its coil has resistance 40Ω.

Determine

(a) the value of the multiplier required to produce a voltmeter reading up to 50V,

(b) the value of the shunt required to convert the instrument to an ammeter reading up to 10A.

(a) Total resistance required to restrict the current to 1.2mA from a 50V supply is

$$R = \frac{50 \text{ V}}{(1.2/1000) \text{ A}} \text{ (note the conversion to amperes)}$$

$$= 41\ 667\Omega$$

The accurate value of additional series resistance required is

R_s = 41 667 – 40

   = 41 427Ω

(b) Voltage required to produce f.s.d. $= \dfrac{1.2}{1000} \times 40$

$= \dfrac{4.8}{1000}$ V

Then $\dfrac{4.8}{1000}$ $V = R_p \times (10A - 0.0012A)$

$\therefore R_p = \dfrac{4.8}{1000 \times 9.9988}$

   $= 4.8 \times 10^{-4}\Omega$

Again, the 1.2mA of current which flows through the instrument could have been neglected in calculating R_p.

EXERCISE 13

1 The coil of a moving-coil instrument has resistance 50Ω, and a current of 0.8mA is required to produce full-scale deflection. Calculate the voltage required to produce full-scale deflection.

2 A moving-coil instrument movement was tested without either shunt or multiplier fitted and it was found that at full-scale deflection the current through the coil was 1.15mA and the voltage across it was 52mV. Determine the resistance of the coil.

3 A moving-coil instrument gives full-scale deflection with a current of 1.5mA and has resistance (without shunt or multiplier) of 25Ω. Determine the value of additional series resistance (the multiplier) required to produce a voltmeter capable of measuring up to 150V.

4 Using the instrument movement of Question 3, modify the metre to measure currents up to 25A by calculating the value of a suitable shunt resistor.

5 Given an instrument movement of resistance 40Ω and requiring a current of 1mA to produce f.s.d., determine the values of the various resistors required to produce a multi-range instrument having the following ranges:

Voltage	0–10V	0–150 V	0–250V
Current	0–1A	0–10A	

6. A moving-coil instrument requires 0.75mA of current to produce f.s.d. at a voltage of 50mV. The resistance of the coil is

    (a) 0.015Ω   (b) 0.067Ω   (c) 66.7Ω   (d) 0.0518V

7 The coil of a moving-coil instrument has resistance 45Ω and requires a current of 1.15mA to produce f.s.d. The p.d. required to produce f.s.d. is

    (a) 51.8V   (b) 39V   (c) 0.025V   (d) 51.8V

8 The value of the multiplier required to convert the instrument of Question 7 to a voltmeter to measure up to 250V is approximately

    (a) 217kΩ   (b) 6.25Ω   (c) 217Ω   (d) 288 kΩ

9 The approximate value of the shunt required to convert the instrument of Question 6 to an ammeter to measure up to 25A is

    (a) 2Ω   (b) 0.002Ω   (c) 0.067Ω   (d) 0.125Ω

Alternating Current Motors

11

For single-phase motor:

Power = Voltage × current × power factor

$P = U \times I \times pf$

For a three-phase motor:

Power = $\sqrt{3}$ × *line voltage × line current × power factor*

$p = \sqrt{3} \times U_L \times I_L \times pf$

EXAMPLE 1

Calculate the current taken by a 1.7 *kW* 230 *volt* single-phase motor working at full load with a p.f. of 0.8 and an efficiency of 75%.

It should be noted that when a motor power rating is given it is the output power unless otherwise stated.

Output is 1.7 *kW* or 1700 *W*.

Calculation for current drawn is:

$P = U \times I \times pf$

$$\frac{p}{U \times pf}$$

$$\frac{Output \times 100}{U \times pf \times efficiency}$$

It is an easier calculation if the efficiency is shown as a decimal and put on bottom line:

$$= \frac{1700}{230 \times 0.8 \times 0.75}$$

Enter into calculator 1700 ÷ (230 × 0.8 × 0.75) = (*ans*)

$$= 12.31A$$

EXAMPLE 2

A three-phase 400V induction motor with an output of 12.4kW is to be installed to drive a conveyor belt. The motor has a p.f. of 0.85 and an efficiency of 78%. It is to be protected by a BS EN 60898 circuit breaker type C. Calculate the current drawn per phase and the size of the protective device.

$$P = \sqrt{3}\, U_L I_L \times pf$$

$$I_L = \frac{P}{\sqrt{3} \times U_L \times pf \times eff}$$

$$= \frac{12400}{\sqrt{3} \times 400 \times 0.85 \times 0.78}$$

(note 78% changed to 0.78 and put on the bottom)

Enter into calculator 12400 ÷ ($\sqrt{3}$ × 400 × 0.85 × 0.78) = (*ans*)

(note the change of % to a decimal)

$$= 26.99A$$

Remember: The protective device must be equal to or greater than the design current (current drawn per phase). Table 41.3 in Chapter 41 of BS 7671 lists the sizes of protective devices.

In this case a 32A device should be used.

EXAMPLE 3

A three-phase 400V induction motor connected in star has an output of 18kW and a p.f. of 0.85. The motor circuit is to be protected by a BS 88 fuse. Calculate (a) the design current (the current drawn from the supply) and (b) the correct rating of the protective device for this circuit.

A. Design current

$$I_L = \frac{P}{\sqrt{3} \times U_L \times pf}$$

$$= \frac{1800}{\sqrt{3} \times 400 \times 0.85}$$

$I_L = 30.56A$

B. Protective device is 32A (if unsure of the ratings of fuses, Table 41.2 in BS 7671 is one table that can be used for fuses.)

EXAMPLE 4

The same motor as in Example 3 is connected in delta. Calculate (a) design current, (b) the output power and (c) the correct size of protected device.

Remember: the output of the motor will increase if it is connected in delta.

A. Design current I_L in delta 30.56 × 3

= 91.7A (3 × *current in star*)

B. Output power

$$P = \sqrt{3} \times U_L \times I_L \times pf$$

$$= \sqrt{3} \times 400 \times 91.7 \times 0.85$$

Output power = 54000 *watts or* 54*kW* (3 × *output in star*)

C. Protective device 100*A*.

Exercises 3 and 4 show by calculation that more current will be drawn from the supply when connected in delta. This is the reason why it is common for three-phase motors to be started in star and then change to delta.

The starting current for motors is considerably greater than their running current, between five and ten times depending on the load being driven.

It can be seen that in delta the start current for the motor will be at least:

30 × 5 = 150*A* when started in star.

53 × 5 = 265*A* when started in delta.

This is why particular care should be taken when selecting protective devices, it is important that a device which can handle medium to high inrush currents is selected.

EXAMPLE 5

A 400V three-phase motor with a p.f. of 0.7 has an output of 3.2kW. Calculate (a) the line current, (b) the power input of of the motor (kVA) and (c) the reactive component of the motor.

$$P = \sqrt{3}\, U_L I_l pf$$

transpose

A. $\dfrac{3200}{\sqrt{3} \times 400 \times 0.7}$

= 6.6 *Amperes*

B. $pf = \dfrac{output}{input}$ or $\dfrac{kW}{kVA}$

Transpose for power factor $kVA = \dfrac{kW}{pf}$

$= \dfrac{3200}{0.7}$

kVA = 4571 *VA or* 4.571kVA

C.$kVAr^2 = kVA^2 - kW^2$

$= \sqrt{kVA^2 - kW^2}$

$= \sqrt{4.571^2 - 3.2^2}$

$kVA\, r$ = 45.59

EXAMPLE 6

A load of 300kg is to be raised through a vertical distance of 12m in 50s by an electric hoist with an efficiency of 80%. Calculate the output required by a motor to perform this task.

Figure 79 A three-phase motor

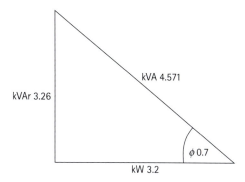

The force required to lift a mass or load of 1kg against the force of gravity is 9.81N.

Work on load = force × distance

12 × 300 × 9.81

= 35316 Newton metres   1 Nm = 1 joule.

Therefore work required to be done to lift the load is 35 316 Joules.

Output required by motor $= \dfrac{\text{Energy in joules}}{\text{Time in seconds}}$

$= \dfrac{35316}{50 \times 0.8}$

(*The efficiency of the hoist is used as a decimal, always under to increase the output*) 882.9 *watts or joules per second.*

EXAMPLE 7

The motor selected for use in Exercise 6 is a single-phase 230V induction motor with an output of 1*kW* and a p.f. of 0.85.

Calculate (a) the current drawn from the supply and (b) the correct size of BS EN 60898 type C protective device.

$P = U \times I \times pf$

(a) $I_l = \dfrac{P}{U_L \times pf}$

(b) $= \dfrac{1000}{230 \times 0.8}$

$= 5.43A$

(c) Protective device chosen would be a 6A type C BS EN 60898.

EXERCISE 14

1. Calculate the full-load current of each of the motors to which the following particulars refer:

	Power output (kw)	Phase	Voltage	Efficiency (%)	Power factor
A	5	1	230	70	0.7
B	3	1	250	68	0.5
C	15	3	400	75	0.8
D	6	1	230	72	0.55
E	30	3	400	78	0.7
F	0.5	1	230	60	0.45
G	8	3	400	65	0.85
H	25	3	400	74	0.75

2. You are required to record the input to a single-phase a.c. motor in kW and in kVA. Make a connection diagram showing the instruments you would use. A 30kW single-phase motor delivers full-load output at 0.75 p.f. If the input is 47.6kVA, calculate the efficiency of the motor.

3. A single-phase motor develops 15kW. The input to the motor is recorded by instruments with readings as follows: 230V, 100A and 17 590W. Calculate the efficiency of the motor and its p.f. Draw a diagram of the connections of the instruments. Account for the energy lost in the motor.

4. Make a diagram showing the connections of a voltmeter, an ammeter and a wattmeter, in a single-phase a.c. circuit supplying power to a motor. The following values were recorded in a load test of a single-phase motor. Calculate the efficiency of the motor and its power factor:

Voltmeter reading 230V
Ammeter reading 75A
Wattmeter reading 13kW
Mechanical output 10kW

5. A single-phase motor drives a pump which raises 500kg of water per minute to the top of a building 12m high. The combined efficiency of the pump and motor is 52%, the supply voltage is 230V and the p.f. is 0.45. Calculate the supply current.

6. The output of a motor is 75kW; the input is 100kW. The efficiency of the motor is

    (a) 13.3%   (b) 7.5%   (c) 1.33%   (d) 75%

7. The efficiency of a motor is 80%. The input power when its output is 24kW is

    (a) 30kW   (b) 19.2kW   (c) 192kW   (d) 300kW

Application of Diversity Factors

12

In the majority of installations, if the rating (I_N) of the protective devices contained within the distribution board were added together, the total current would, in most cases, exceed the rating of the supply fuse, often quite considerably. Surprisingly, the main fuse very rarely 'blows'.

If we consider a typical installation it is quite easy to see why this scenario exists.

It is very unusual in a domestic situation for all of the installed loads to be in use continuously. To minimise the size of cables and other equipment we may apply diversity factors to certain circuits to obtain an assumed maximum demand for an installation.

Appendix A of the *IET On-site Guide* gives information on the calculation of assumed current demand and allowances for final circuit diversity.

It should be remembered that diversity is not a precise calculation and in many instances experience and knowledge could be used instead of the tables which are a good guide only.

EXAMPLE 1

A domestic premises with a 230V 50H$_z$ supply protected by a 100A BS 1361 supply fuse has the following loads.

Shower: 10 kW protected by 45A device

Cooker: 12 kW (13A socket outlet in control unit) 45A device

2 × 32 amp ring final circuits

1 × 20 amp radial circuit serving socket outlets

1 × 16 amp immersion heater circuit

2 lighting circuits, each with 10 outlets (min 100*watt* per lamp Table A1 *IET On-site Guide*) 2kW protected by 2no. 6A devices.

If we total up the current rating of the protective devices it can be seen that we have a potential current of:

45 + 45 + 32 + 32 + 20 + 16 + 6 + 6 = 202 Amperes

If we use Table A2 from the *IET On-Site Guide* to apply diversity factors we can find the assumed current.

$$\frac{10000}{230} = 43.47A$$

Cooker. Diversity allowed see Table A2 row 3

$$\frac{12000}{230} = 52A$$

From Table A2 use first 10A and 30% of remainder which is 42 × 30% = 12.6*A*

Enter on calculator 42 × 30% = (*Ans* 12.6)

Total of 10 + 12.6 = 22.6 Add 5*A* for socket outlet 27.6*A*

Power circuits. Row 9 of Table A2

2 × 32*A and* 1 × 20*A* Standard arrangement circuit.

32A plus 40% of 52A (52 × 40% = 20.8A) Total 52.8A

Immersion heater. Row 6 of Table A2

No diversity. 16*A*

Lighting. Row 1 of Table A2, 66% is permitted

$$\frac{2000}{230} = 8.69 \times 66\% = (Ans\ 5.73A)$$

Enter on calculator 8.69 × 0.66 =

Total assumed demand is now:

43 + 27 + 52 + 16 + 5 = 143A

In this case as in many other cases, the assumed maximum demand is still greater than the supply fuse. You will find that the supply fuse has been in place for many years and never been a problem and should not give rise for concern. This is because the diversity allowed is usually quite a conservative figure.

The biggest problem with this situation is that the switched disconnector is usually 100A. In these cases consideration should be given to using a split load board with possibly the power circuits on an RCD protected side of the board.

EXAMPLE 2

A retail jewellers has the following connected load supplied at 230V 50Hz:

Direct heaters 2 × 2kW,   3 × 1.5kW,   1 × 1.0kW
Cooker 5kW (cooker unit has socket outlet)
Water heating (thermostatic) 3kW
Socket outlets 30A ring circuit
Shop and window lighting 2.5kW total.
Determine the assumed maximum demand.

Assumed maximum demand (A.M.D.) using *IET On-Site Guide* Table A2 (small shop premises):

Heaters = 2 × 2kW + 3 × 1.5 + 1.0

$$= \frac{2000}{230} + \frac{7500}{230} \times \frac{75}{100}$$

$$= 8.7A + 24.46A = 33.16A$$

$$\text{Cooker} = \frac{5000}{230} = 21.7A$$

Socket at cooker control 5A

$$\text{Water heaters} = \frac{3000}{230} = 13.04A \text{ (no diversity)}$$

Socket outlets 30A

$$\text{Lighting} = \frac{2500}{230} = 10.9 \times \frac{90}{100} = 9.8A$$

A.M.D. = 33.16 + 21.7 + 5 + 13.04 + 30 + 9.8 = 112.7A

In this case there may be a separate main control for associated circuits. Heating and shop window lighting may be on time switch/contactor controls with individual distribution boards and switch-disconnectors.

EXAMPLE 3

A small hotel supplied at 400/230V 50Hz has the following connected load:

100 lighting outlets
50 × 13A socket outlets on 6 × 30A ring circuits
30 × 1kW convection heaters on coin-operated meters
16 k W thermal storage central heating boiler
Cooking equipment: 2 × 14kW cookers, 1 × 4kW hot cupboard, 3 × 2kW fryers,
  4 × 600W microwave ovens plus 5kW machines.
Determine the assumed maximum demand.

Using Tables A1 and A2 in the *IET On-Site Guide:*

100 lighting points = 100 × 100W = 10kW so A.M.D. = $\dfrac{10\ 000}{230} \times \dfrac{75}{100} = 32.6A$

6 × 30A ring circuits: A.M.D. is $30A + \dfrac{150 \times 50}{100} = 105A$

30 × 1kW convection heater: A.M.D. is $\dfrac{1000}{230} + \dfrac{29000 \times 50}{230 \times 100} = 4.35A + 63A = 67.35A$

16kW thermal storage: A.M.D. is $\dfrac{16000}{230} = 69.56A$ (no diversity)

Cooking equipment: $\dfrac{14000}{230} + \dfrac{14000 \times 80}{230 \times 100} + \dfrac{4000 + 6000 + 2400 + 5000 + 60}{230 \times 100}$

= 60.9A + 48.7A + 45.4A = 155A

A.M.D.

= 32.6A + 105A + 67.35A + 69.56A + 155A

= 429.51A

Assuming that load is balanced over the three-phase supply then load would be approximately 143A per phase.

EXERCISE 15

1 A bungalow is supplied at 230V 50Hz and has the following connected load:

   18 ceiling mounted lighting outlets

   12 × 2A socket outlets for local luminaires

   3 × 30A socket outlet ring circuits

   1 × 10kW cooker (control unit without socket outlet)

   1 × 5.5kW hob unit 10kW of thermal storage space heating

   1 × 3kW immersion heater (thermostat controlled)

   1 × 8kW shower unit.

Determine the assumed maximum demand, and comment upon the magnitude of this.

2 A ladies' hairdressing salon is supplied at 230V 50Hz and has the following connected load:

   4kW thermal storage space heating

   6 × 3kW under-sink instantaneous water heaters

   2 × 30A socket outlet ring circuits

   2kW of shop lighting

   2 × 500W tungsten-halogen shop front luminaires

Determine the assumed maximum demand and comment upon its magnitude.

3 A country hotel is at present supplied at 230V 50Hz and is to be rewired employing the following installation circuits:

   Lighting: four floors each having 1000W loading

   Heating: three upper floors each having 6 × 1kW convection heaters; ground floor 3 × 3kW and 3 × 2kW convection heaters

   Socket outlets: 4 × 30A ring circuits

Cooking appliances: 1 × 10kW cooker, 1 × 6kW hob unit, 4kW of assorted appliances (cooker control without socket outlet)

Outside lighting 3 × 500W tungsten halogen floodlights

Determine the assumed maximum demand and comment upon its magnitude.

4 An insurance office is supplied at 400/230V 50Hz and has the following connected load:

4 × 30A socket outlet circuits for computer use

1 × 30A socket outlet circuit for general use

1.5kW of fluorescent lighting

1.0kW of tungsten lighting

1 × 6kW cooker

2 × 600W microwave cookers

2 × 3kW instantaneous type hand washers

2 × 2kW hand dryers.

Determine the assumed maximum demand.

Cable Selection

13

EARTHING CONDUCTOR CALCULATION

Table 54.7 in BS 7671 gives the minimum size for circuit protective conductors. This is a simple table to use, however it is often useful to use a smaller c.p.c. than that required by Table 54.7. On larger cables cost is a major factor as is space. If it is required to select a size of conductor smaller than is given in the table, a calculation must be carried out to ensure the conductor temperature will not rise above its final limiting temperature under fault conditions, this is called the adiabatic equation (final limiting temperature for conductors can be found in Table 43.1).

The line and c.p.c. must also meet Z_s requirements for the circuit.

BS 1361 protective devices are now listed in the IET wiring regulations as BS 88–3 first system 3.

EXAMPLE 1

The design current (Ib) for a circuit is 38A.

The current-carrying capacity of cable has been calculated and the circuit is to be wired in 70°C thermoplastic singles, 6mm^2 live conductors and 1.5mm^2 c.p.c. It is protected by a BS 60898 40A Type B circuit breaker. Supply is 230V TNS with a Z_e of 0.38 Ω, circuit is 28m long.

Calculate Z_s and requirements and thermal constraint.

The resistance of the line and c.p.c. for this circuit must now be calculated and then compared with the values given in Table 41.3 in BS 7671.

From Table I1 in the *IET On-Site Guide* it will be seen that the $r_1 + r_2$ for $6mm^2/1.5^2$ copper is $15.2m\Omega$ per metre @ a temperature of 20°C.

The cable resistance is given in Table I1 is at 20°C and the Z_s for protective devices given in tables in BS 7671 is for cables at their operating temperature of 70°C.

The cable resistance must be adjusted by calculation to allow for the increase in resistance due to the rise in temperature. The resistance of the copper conductor will increase 2% for each 50°C rise in temperature.

If a cable temperature alters from 20°C to 70°C the resistance will rise 20%.

Table I1 from the *IET On-Site Guide* gives multipliers to correct the resistance of conductors at the maximum operating temperature depending on the type of cable insulation and how the c.p.c. is installed.

For 70°C thermoplastic (pvc) multicore cable it can be seen that a multiplier of 1.20 must be used (if you multiply by 1.20 a value will increase by 20%).

$$\frac{m\Omega \times L \times 1.2}{1000} = R_1 + R_2$$

$$\frac{15.2 \times 28 \times 1.2}{1000} = 0.511\Omega$$

$Z_s = Z_e + (R1 + R2)$ or $Z_s = 0.38 + 0.51 = 0.89\Omega$ (two decimal places is ok)

$Z_s = 0.89\ \Omega$

Compare this with value, the maximum value given for Z_s in Table 41.3. The calculated value must be less than the tabulated value.

Max Z_s for 40A type B mcb is 1.2Ω therefore our circuit Z_s at 0.89Ω is acceptable.

Now the prospective earth fault current must be calculated.

The calculation is:

$$\frac{U_0 C}{Z_s} = I_f \qquad \frac{230}{0.89} = 258.4A$$

Prospective earth fault current is 258.4A (0.258 *kA*)

The calculation for thermal constraint should now be carried out.

From Figure 3A4 in Appendix 3 of BS7671 the disconnection time for a 40A BS 60898 type B device can be found.

From the chart on the top right of the page it can be seen that for a 40A device to operate within 0.1s a minimum current of 200A is required.

Regulation 543.1.3 gives the formula that must be used for thermal constraint.

The adiabatic equation is:

$$S = \frac{\sqrt{I^2 \times t}}{k}$$

S is the minimum permissable size of c.p.c.

I is the earth fault current

t is the time in seconds

k is the value given from Table 54C for 70°C cable

$$\frac{\sqrt{258 \times 258 \times 0.1}}{115} = 0.71\text{mm}^2$$

Enter into calculator 258 x² 0.1 = $\sqrt{}$ = ÷ 115 = answer (0.71)

The minimum size c.p.c that may be used is 0.73mm².

Therefore 1.5mm² cable is ok.

EXAMPLE 2

A single-phase 230V circuit is to be wired in 10mm² phase with 1.5mm² c.p.c. thermoplastic 70°C copper singles cable. The protective device is a 63A BS 88 general purpose fuse. Z_e for the circuit is 0.23Ω. The circuit is 36m long, and has a maximum of 5s disconnection time.

Calculate to find actual Z_s and for thermal constraints.

From Table I1 of the *IET On-Site Guide* 10mm² (r1) has a resistance of $\frac{1.83m\Omega/m}{m}$

and 1.5mm² (r2) has a resistance of $\frac{12.10m\Omega/m}{m}$

Resistance of cable is 1.83 + 12.10 = 13.93$m\Omega$ per metre. Therefore: 36m will have a resistance of:

$$\frac{13.93 \times 36 \times 1.2}{1000} = 0.6\Omega$$

(*remember the multiplier 1.2 for temperature correction*)

Calculate actual Z_s

$Z_s = Z_e + (R_1 + R_2)$ or $0.23 + 0.6 = 0.83\Omega$

$Z_s = 0.83\Omega$

Compare this value with maximum permissible Z, from 41.4 in BS 7671.

Maximum permissible Z_s for a 63A BS 88 is 0.86Ω. This will be fine as the calculated Z_s is 0.83Ω.

Now we must calculate maximum earth fault current:

$$\frac{U°C}{Z_s} = I_f$$

$$\frac{230}{0.83} = 277.1A$$

Now use I_f to calculate disconnection time using Figure 3A3(b) in BS 7671 as follows.

Along the bottom line move to the right until a vertical line matching a current of 290A is found, follow the line up vertically until it crosses the thick black line for the 63A fuse, in line with this junction move across to the left hand side to find the disconnection time, it will be 4s.

Now look in Table 54.3 of BS 7361 to find the value K for the protective conductor.

A 70°C thermoplastic cable with a copper conductor has a K value of 115.

Now carry out the adiabatic equation to ensure that c.p.c. is large enough.

On calculator enter $289.912 \times 4 \sqrt{} = \div 115 = (5.04)$

$$\frac{\sqrt{I^2 \times t}}{115} = S$$

$$\frac{\sqrt{289.91 \times 289.91 \times 4}}{114} = 5.05mm^2$$

This shows that the c.p.c. is too small

The same calculation should now be carried out using 2.5mm^2 c.p.c.

From Table l1 of the *IET On-Site Guide* 2.5mm^2 has a resistance of

10mm^2/25mm^2 has a resistance of 1.83 + 7.41 = = $\dfrac{9.24m\Omega/m}{m}$

7.41$\dfrac{m\Omega}{m}$ $\dfrac{9.24 \times 36 \times 1.2}{1000} = 0.399\Omega$

$I_f = \dfrac{230}{0.399} = 576.44$

Now check disconnection time in Figure 3A3(b) of Table 3 BS 7671.

Disconnection time is now 0.2s.

Use the adiabatic equation to calculate the minimum size c.p.c. that could be used under the conditions which it is to be installed:

$\dfrac{\sqrt{601.5^2 \times 0.2}}{115} = 2.33$mm^2

This proves 2.5mm^2 c.p.c. can be used.

EXERCISE 16

1 Calculate the R$_1$ + R$_2$ of 23m of the copper conductors in a 2.5mm^2 /1.5mm^2 thermoplastic twin and c.p.c. cable at 20°C.

2 Calculate the resistance of the conductors in Question 1 at their operating temperature of 70°C.

3 A circuit is to be wired in 70°C thermoplastic 6mm^2 /2.5mm^2 copper cable and is 18m long, the Z$_e$ for the circuit is 0.8Ω. Calculate the Z$_s$ for the circuit at its maximum operating temperature.

4 The circuit above is protected by a BS 3036 semi-enclosed fuse with a disconnection time of 5s. Will the circuit comply with the requirements of BS 7671?

5 A circuit is wired in thermoplastic copper 4mm^2 phase with a 1.5mm^2 c.p.c. It has a calculated Z$_s$ of 1.14Ω at 70°C. The circuit is protected by a 30A BS 1361 fuse with a maximum disconnection time of 0.4s. Will this cable comply with the requirements for the required disconnection time and thermal constraints?

6 If a circuit was wired in 90°C thermosetting cable with copper conductors, and had a calculated fault current of 645A with a disconnection time of 1.5s, calculate using the adiabatic equation the smallest permissible size c.p.c.

7 A circuit is required to supply a 60A load, it is to be installed in trunking using 70°C thermoplastic (pvc) single cables with copper conductors. The circuit will be protected by a BS 88 fuse, the trunking will contain two other circuits and will be fixed using saddles to a brick wall in an ambient temperature of 35°C. Maximum permissible voltage drop for this circuit is 6V. The circuit is 27m long, supply is 230V TNC-S with a Z_e of 0.35Ω. Disconnection time is 5s maximum.

Calculate the minimum size phase and c.p.c. conductors required.

VOLTAGE DROP CALCULATIONS AND CABLE SELECTION CALCULATIONS

The calculations are fully explained in book 1. These are additional questions for revision.

Voltage drop calculations

The voltage drop in cable conductor(s) is directly proportional to the circuit current and the length of cable run.

$$\text{Voltage drop} = \frac{\text{current(A)} \times \text{length of run (m)} \times \text{millivoltage drop per A/m}}{1000}$$

Note division by 1000 to convert millivolts to volts.

Note BS 7671 Requirement 525.1 limits the voltage drop permitted between the origin of the supply and the load to 3% for lighting and 5% for all other circuits. For a 230V single-phase supply this is 6.9V for lighting and 11.5V for other circuits.

EXAMPLE 1

A 3kW 230V 50Hz single-phase motor has an efficiency of 70% and works at a p.f. of 0.6. It is connected to its starter by single-core pvc-insulated cables (copper conductors) drawn into steel conduit (method B); the length of run is 25m. The voltage drop in the cables must not exceed 6V. Assume an ambient temperature of 35°C and protection by BS 88 fuses.

Circuit details:

Motor circuit, starter will offer overload protection.
Ambient temperature 35°C *so* C_a is 0.94.
Using BS 88 (Gm) fuses so C_r is 1.
Output = 3kW

$$\text{Input} = 3000 \times \frac{100}{70}$$
$$= 4285.7\text{W}$$

$P = U \times I \times \text{p.f.}$

$$4285.7 = 230 \times I \times 0.6$$
$$I_b = \frac{4285.7}{230 \times 0.6}$$
$$= 31.1\text{A}$$

Minimum BS 88 fuse rating (I_n) say 40A (allows for moderate overcurrent at starting). Starter will offer overload protection (see BS 7671 Requirements 435.2 and 552.1.2).

Rating factors applying:

C_a is a 0.94 (35°C)
C_r is 1 (BS 88 fuses)

Thus minimum current rating:

$$(I_t) = \frac{40}{0.94 \times 1}$$
$$= 42.55\text{A}$$

Using BS 7671 Table 4D1A or *IET On-Site Guide* Table F4i, from column 4 select 10mm^2 cables (57A) and using BS 7671 Table 4D1B or *IET On-Site Guide* Table F4ii, column 3, read mV/A/m value for 10mm^2 cables as 4.4mV/A/m

$$\text{Voltage drop in 25m} = \frac{31.1 \times 25 \times 4.4}{1000}$$
$$= 3.42\text{V}$$

Thus 10mm^2 cables will be suitable.

EXAMPLE 2

(a) An industrial process heater of rating 16kW is fed at 400V 50Hz. Three-phase four-wire is to be installed in a factory using a pvc-insulated, non-armoured, copper conductors multicore cable. Length of run is 25m clipped direct to a wall; assume a maximum ambient temperature of 35°C and protection by BS 3036 fuses.

(b) If the BS 3036 fuses were replaced by BS 88 (Gg) fuses what would be the effect on cable current rating?

Circuit details:

As it is a heater p.f. is unity.
Ambient temperature 35°C so C_a is 0.94.
Using BS 3036 fuses so C_c is 0.725.

Current demand $I_b = \dfrac{16000}{\sqrt{3 \times 400}}$

$\qquad\qquad\qquad = 23.1\text{A}$

(a) Select as I_n 30A BS 3036 fuses. Thus minimum current rating is

$I_t = \dfrac{30}{0.94 \times 0.725}$

$\quad = 44\text{A}$

Using BS 7671 Table 4D2A or *IET On-Site Guide* Table F5i, from column 7 select 10mm^2 cables (57A) and using BS 7671 Table 4D2B or *IET On-Site Guide* Table F5ii, column 4, read mV/A/m value for 10mm^2 cables as 3.8 mV/A/m.

Voltage drop in 25m $= \dfrac{23.1 \times 25 \times 3.8}{1000}$

$\qquad\qquad\qquad\quad = 2.19\text{V}$

Thus 10mm^2 cables will be suitable.

(b) Select as I_n 25A BS 88 fuses. In this case C_c is 1. Thus minimum current rating is

$$I_t = \frac{25}{0.94 \times 1}$$

$$= 26.6A$$

Using BS 7671 Table 4D2A or *IET On-Site Guide* Table F5i, from column 7 select 4mm^2 cables (32A) and using BS 7671 Table 4D2B or *IET On-Site Guide* Table F5ii, column 4, read mV/A/m value for 4mm^2 cables as 9.5 mV/A/m.

$$\text{Voltage drop in 25m} = 25m = \frac{23.1 \times 25 \times 9.5}{1000}$$

$$= 5.49V$$

Thus 4mm^2 cables will be suitable.

EXAMPLE 3

A pvc trunking containing single-core pvc-insulated distribution cables (copper conductors) is to be run 30m from the main switchgear of an office building to supply a new 400/230V T.P. & N. distribution fuseboard. The balanced load consists of 18kW of discharge lighting. The main and local distribution boards employ fuses to BS 88 (Gg) Part 2. The voltage drop in the distribution cables must not exceed 6V and the ambient temperature is anticipated to be 30°C. For the distribution cables, establish and state the

(i) design current I_b
(ii) minimum rating of fuse in the main switch fuse I_n
(iii) maximum mV/A/m value
(iv) minimum current rating I_t
(v) minimum cross-sectional area of the live conductors
(vi) actual voltage drop in the cable.

Circuit details:

Discharge lighting circuit requires a multiplier of 1.8
(*IET On-Site Guide*, Appendix A)
Ambient temperature 30°C so C_a is 1.
Using BS 88 fuses so C_c is 1.
Cable voltage drop limitation of 6V.
Cables in trunking to method B.

(i) Design current $I_b = \dfrac{18 \times 10^3 \times 1.8}{\sqrt{3} \times 400}$

$$= 46.77\text{A}$$

(ii) Minimum BS 88 fuse rating is 50A.

(iii) Maximum mV/A/m value $= \dfrac{6 \times 1000}{46.77 \times 30}$

$$= 4.28\text{mV/A/m}$$

(iv) Minimum current rating $= \dfrac{50}{1}$

(v) From BS 7671 Tables 4D1A and 4D1B or *IET On-Site Guide* Table F4i (column 5) and Table F4ii (column 6), minimum c.s.a. of cable is 16mm^2 (68 A/2.4 mV/A/m).

(vi) Actual voltage drop in 30m cable $= \dfrac{46.77 \times 33 \times 2.4}{1000}$

$$= 3.7\text{V}$$

EXAMPLE 4

A 400V 50Hz three-phase extractor fan has a rating of 15kW at 0.8 p.f. lagging and is supplied from a BS 88 (Gg) Part 2 type distribution board 40m distant. The cables are to be single-core, pvc-insulated, run in steel trunking with three similar circuits. Assume an ambient temperature of 35°C and that the voltage drop in the cables is limited to 2.5% of the line voltage. Establish the:

(i) full load current of the motor I_L
(ii) rating of the fuses I_n
(iii) minimum current rating of cables
(iv) minimum cable c.s.a.
(v) actual voltage drop in cables.

Circuit details:

Extract fan circuit: low starting current.
Four sets of circuit cables: C_g is 0.65.
Ambient temperature: 35°C so C_a is 0.94.
Using BS 88 fuses, so C_c is 1.
Cable voltage drop limitation of 2.5% of 400, i.e. 10V.
Cables in trunking to method B.

(i) As $P = \sqrt{3}U_L I_L \cos\phi$

$15\ 000 = \sqrt{3} \times 400 \times I_L \times 0.8$

$I_L = 1500\ \sqrt{3} \times 400 \times 0.8$

$= 27A$

(ii) Select 32A BS 88 fuses (allowing for low starting current).

(iii) Minimum current rating of cable $= \dfrac{32}{0.94 \times 0.65}$

$= 52.4A$

(iv) From BS 7671 Tables 4D1A and 4D1B or *IET On-Site Guide* Table F4i (column 5) and Table F4ii (column 6) select 16mm^2 (68 A/2.4 mV/A/m).

(v) Voltage drop in 40m $= \dfrac{27 \times 40 \times 2.4}{1000}$

$= 2.6V$

As volts drop limitation is 2.5% of 400V, i.e. 10V, 16mm^2 cable is satisfactory. The final example illustrates the effect on the required tabulated cable rating of combined rating factors.

EXAMPLE 5

A twin and earth pvc-insulated (copper conductor) cable runs between a 230V distribution board at the origin of an installation and a 10kW heater. The cable passes through the following environmental conditions:

(a) on its own in a switchroom with an ambient temperature of 35°C;
(b) on its own in an outdoor area with an ambient temperature of 25°C;
(c) bunched with three other cables on a wall surface in an area with an ambient temperature of 40°C;
(d) finally on its own passing through a thermally insulated wall section for a distance of 2m, in an ambient temperature of 30°C.

Protection is by BS 3036 fuses, length of run is 60m and the voltage drop is limited to 5.5V.

Calculate the minimum cable rating and select suitable cable for voltage drop limitation.

Circuit details:

Heater circuit so no special restrictions.

Protection by BS 3036 fuses so C_c is 0.725.

Voltage drop limitation is 5.5V.

Area (a) 35°C C_a is 0.94.

Area (b) 25°C C_a is 1.03.

Area (c) 40°C C_a is 0.87, C_g is 0.65.

Area (d) 30°C C_a is 1, C_i is 0.5.

Now overall rating factors are as follows:

Area (a) $0.94 \times 0.725 = 0.68$

Area (b) $1.03 \times 0.725 = 0.747$

Area (c) $0.87 \times 0.65 \times 0.725 = 0.41$

Area (d) $0.5 \times 0.725 = 0.36$ (worst area)

Design current $I_b = \dfrac{10000}{230}$

$\qquad\qquad = 43.5A$

Nearest BS 3036 fuse element is 45A (BS 7671 Table 52.2A). Select worst area (d): C_i is 0.5.

Minimum cable rating $= \dfrac{45}{0.36}$

$\qquad\qquad\qquad = 125A$

From BS 7671 Table 4D2A or *IET On-Site Guide* Table F5i and from BS 7671 Table 4D2B or *IET On-Site Guide* Table F5ii select 70mm^2 (139A) and 0.63 mV/A/m.

Voltage drop in 60m $= \dfrac{43.5 \times 60 \times 0.63}{1000}$

$\qquad\qquad\qquad\quad = 1.64V$

So 70mm^2 cable is satisfactory.

Obviously one should avoid running cables in hostile environments wherever possible, in this case avoiding thermal insulation and not using BS 3036 protection. Assuming that the cable grouping was unavoidable we could now use area (c) as the worst environment and in this case:

Revised circuit details:

Heater circuit so no special restrictions.
Protection by BS 88 fuses so C_c is 1.
Voltage drop limitation is 5.5V.
Area (c) $0.87 \times 0.65 = 0.565$
Voltage drop limitation 5.5V.

Minimum cable rating $= \dfrac{45}{0.565}$

$= 79.6\text{A}$

From BS 7671 Table 4D2A or *IET On-Site Guide* Table F5i and from BS 7671 Table 4D2B or *IET On-Site Guide* Table F5ii select 25mm^2 (90A) and 1.75 mV/A/m.

Voltage drop in 60m $= \dfrac{43.5 \times 60 \times 1.75}{1000}$

$= 4.57\text{V}$

So 25mm^2 cable is satisfactory and cheaper to install than 70mm^2 cable.

EXERCISE 17

1 Establish the current-carrying capacity (I_t) of a 70° pvc insulated cable with a tabulated current rating (I_t) of 17.5A when it is grouped in conduit with two other circuits in an ambient temperature of 35°C; protection is by BS 3036 fuses.

2 Calculate the actual voltage drop and the power wasted in a 25mm^2 cable, 10m long, when it carries 70A. The listed mV/A/m for the cable is 1.8mV.

3 The design current of a single-phase circuit is 35A. The single-core pvc-insulated cables run alone in pvc conduit for a distance of 50m through an area having an ambient temperature of 35°C (100mm of the conduit passes through thermal insulation). The voltage drop in the circuit must not exceed 5V. Protection is by a BS 1361 fuse. Determine the:

    (a) fuse rating

    (b) minimum cable current rating

    (c) minimum cable c.s.a.

    (d) voltage drop in the cables.

4 A supply is required to a 3kW heater which is 25m from a local BS 1361 distribution board. The building is fed at 230V 50Hz single-phase. It is proposed to employ a 2.5mm^2 two-core and earth pvc-insulated (copper conductors) cable for this circuit installed as method 1. Allowing for a 2V drop in the cables feeding the distribution board, determine the:

(a) design current

(b) maximum voltage drop permitted

(c) voltage drop in the cable

(d) actual voltage at the heater.

5 A 10kW motor having an efficiency of 60% is fed from a 220V d.c. supply through cables 20m long and having a listed voltage drop figure of 1.3mV/A/m. Determine the:

(a) design current

(b) voltage drop in the cables when the motor is fully loaded.

6 After the application of correction factors, a pair of single-core pvc-insulated cables in conduit are required to carry 25A from a distribution board to a load 90m away. The voltage drop in the cables should not exceed 5V. Using BS 7671 documents:

(a) calculate the maximum mV/A/m value

(b) select a suitable cable c.s.a.

(c) calculate the voltage drop in the cables.

7 A 12kW load is to be supplied from a 230V main switch-fuse 65m distant. The voltage drop is to be limited to 2.5% of the supply voltage. Overload protection is to be provided by a BS 3036 semi-enclosed fuse. The single-core pvc-insulated cables run in conduit with one other single-phase circuit. Assuming an ambient temperature of 25°C, determine with the aid of BS 7671 documents:

(a) the design current

(b) the fuse rating

(c) the minimum cable current rating

(d) the maximum mV/A/m value

(e) the selection of a suitable cable c.s.a.

(f) the voltage drop in the circuit.

8 A single-phase load of 10kW is to be supplied from a 230V distribution board 120m distant. Overload protection is to be by BS 88 (Gg) Part 2 fuses. The twin with earth, pvc-insulated cable is clipped with three similar cables as BS 7671 method 1 in an ambient temperature of 25°C. Voltage drop in the cables should not exceed 5V. Determine with the aid of BS 7671 documents the:

(a) design current

(b) fuse rating

(c) minimum cable current rating

(d) maximum mV/A/m value

(e) minimum cable c.s.a.

(f) voltage drop in the cables.

9 A 400/230V 50Hz T.P. & N. distribution board is to be installed in a factory to feed 11kW of mercury vapour lighting. Due to the adverse environmental conditions, it is intended to use pvc conduit to contain the single-core pvc-insulated cables (copper conductors). The total length of the run from the main switchboard is 50m. To provide earthing protection it is intended to draw a 4mm² single-core pvc-insulated cable (copper conductors) into the conduit. The following details apply to the installation:

(i) an ambient temperature of 35°C

(ii) BS 88 (Gg) Part 2 fuse protection throughout

(iii) voltage drop in the cables must not exceed 8.5V

(iv) the BS 88 fuses satisfy the requirements of BS 7671 regulation 434.

Establish the:

    (a) design current

    (b) rating of fuses in the main switchboard

    (c) minimum current rating of live conductors

    (d) maximum mV/A/m value of live conductors

    (e) minimum cross-sectional area of live conductors

    (f) actual voltage drop in submain cables

    (g) size of pvc conduit to be used, assuming that one section of the run involves one right-angle bend in 8m.

10 A single steel trunking is to be run from a 400/230V 50Hz main switchboard to feed three S.P. & N. lighting distribution boards containing Type 2 BS 3871 miniature circuit breakers, sited at 5, 12 and 20m distances. Each distribution board feeds 5kW of mercury vapour lighting. The following details apply to the installation:

    (i) ambient temperature in the area is 25°C

    (ii) protection at the main switchboard is by BS 88 fuses

    (iii) single-core pvc-insulated (copper conductors) cables are to be employed

    (iv) voltage drop in the distribution cables must not be greater than 3.5V.

Establish the:

    (a) design current

    (b) maximum mV/A/m value permitted

    (c) fuse rating at the main switchboard

    (d) minimum cable current rating

    (e) minimum cross-sectional area of the distribution cables

    (f) voltage at each distribution board.

11 A 12m length of two-core and earth, pvc-insulated cable is clipped to a surface as BS 7671 method 1. The cable feeds a load of 4kW at 230V 50Hz a.c. The following details apply to the installation:

   (i) p.f. of the load is 0.8 lagging

   (ii) ambient temperature of 20°C

   (iii) protection by a BS 88 (Gg) Part 2 fuse

   (iv) cable voltage drop not to exceed 5V.

Determine the:

   (a) design current

   (b) rating of the fuse

   (c) minimum cable current rating

   (d) maximum mV/A/m value

   (e) minimum cable cross-sectional area

   (f) actual voltage drop in the cable at full load.

12 A 4.5kW single-phase load in a factory is to be supplied from the 400/230V 50Hz suppliers' main switchboard 40m distant, using two-core and earth, pvc-insulated cable. The p.f. of the load is 0.7 lagging and the cable route is through an ambient temperature of 30°C. Protection is by BS 88 (Gg) Part 2 fuses.

Determine the:

   (a) design current

   (b) permissible voltage drop in circuit

   (c) minimum fuse rating

   (d) minimum cable current rating

   (e) maximum mV/A/m value

   (f) minimum cable cross-sectional area

   (g) actual voltage drop in the cable at full load.

13 A 230V 50Hz 8kW electric shower unit is to be installed in an industrial premises using a two-core and earth, pvc-insulated cable, 20m in length. The ambient temperature is 30°C. Protection is by a BS 1361 fuse in a distribution board, the cable volts drop should not exceed 2V.

Determine the:

    (a) design current I_b

    (b) fuse rating I_n

    (c) required cable current rating I_t

    (d) required cable c.s.a.

    (e) actual volts drop in the cable.

14 A 25kW 400V 50Hz three-phase motor operates at 0.85 p.f. lagging on full load. The pvc-insulated single-core cables run together for a distance of 10m with two similar circuits through a trunking to a circuit breaker distribution board. Assume that the circuit breaker is selected to have an operating value of not less than 1.5 times the motor full-load current, the ambient temperature is 35°C and the voltage drop in the cables should not exceed 10V.

Determine the:

    (a) design current

    (b) setting of circuit breaker

    (c) minimum cable current rating

    (d) maximum mV/A/m value

    (e) minimum cable c.s.a.

    (f) actual volts drop in the cable.

15 The voltage-drop figure for a certain cable is 2.8 mV/A/m. The actual volt drop in a 50m run of cable when carrying 45A is:

    (a) 1.2V   (b) 6.3V   (c) 0.1V   (d) 10V

16 The voltage drop allowed in a certain circuit is 6V. The length of run is 35m. The cable used has a voltage-drop figure of 7.3mV/A/m. Ignoring any correction factors, the maximum current which the cable can carry is:

(a) 15A   (b) 23.5A   (c) 41A   (d) 43.8A

17 A circuit is given overload protection by a 30A BS 3036 fuse. The grouping factor C_g is 0.65 and the ambient temperature factor is 0.87. The minimum current-carrying capacity of the cable should be:

(a) 73.2A   (b) 53A   (c) 30A   (d) 41.3A

18 A 10kW 230V a.c. motor operates at 0.75 lagging. The starter offering overload protection is set at 1.5 times the F.L.C. of the motor. Ignoring any correction factors, the minimum current-carrying capacity of the cable to the motor required is:

(a) 43.5A   (b) 58A   (c) 87A   (d) 108.7A

19 A certain cable having a tabulated current rating (I_t) of 18A has correction factors of 1.04, 0.79 and 0.725 applied to compensate for its operating conditions. The operational current rating (I_z) for the cable is:

(a) 30.22A   (b) 24.83A   (c) 13.572A   (d) 10.72A

EARTH LEAKAGE PROTECTION CALCULATIONS

To prevent danger to persons, livestock and property every installation must be protected against the persistence of earth leakage currents. This is generally achieved by providing a low-impedance earth-current leakage path from the installation to the source of supply, i.e. the local distribution transformer.

The leakage path must have a low enough impedance (Z_s) to allow adequate earth leakage current to flow to 'blow' the circuit fuse or operate the circuit breaker and isolate the faulty circuit within a specified time, usually either 5s or 0.4s. BS 7671 gives guidance to the permissible earth-loop impedance values to meet the disconnection times and that document and the *IET On-Site Guide* contain tables which list types of protective device and specify the maximum measured earth fault loop impedance in ohms for each rating of the specific device. Where precise disconnection times are demanded then BS 7671 Appendix 3 contains characteristic curves for fuses and circuit breakers.

Part of the earth leakage path is outside the control of an electricity consumer and its impedance (Z_e) contributes to the total value of earth loop impedance. The value of this external impedance is generally declared by the supplier and is used in the calculation of the 'prospective' Z_s. The declared value of Z_e, however, can never be a precise value because of the supplier's service conditions at the moment of earth fault; thus the actual value of Z_s must always be measured by earth loop impedance test instruments at various points within an installation when the particular circuit is energized and is under test-load conditions.

For the estimation of prospective earth-loop impedance values we may, however, regard Z_e as an empirical or estimated value when assessing the permitted value of the installation's internal cable impedance (or resistance) value.

The internal cable 'impedance' will be determined by the cross-sectional area and resistance (R_1) of the circuit's line conductor and that of the circuit's protective conductor (R_2) from the origin of the installation to the point of connection to current-using equipment when the circuit is energized and the cables are working in their maximum operating temperature.

To predict the actual disconnection time for an earth leakage fault condition we may employ characteristic curves of the protective devices, i.e. fuses and circuit breakers. Appendix 3 of BS 7671 gives specimens of such curves.

Note: For all the following examples and exercises pvc-insulated copper conductors are to be employed.

EXAMPLE 1

An installation is being carried out and it is necessary to estimate the prospective total earth loop impedance of circuits. In order to arrive at a typical value, a lighting circuit is chosen as that is likely to have a fairly high impedance value. The circuit is to be wired in 1.5mm² twin and earth cable (assume a 1.0mm² protective conductor); the length of cable is 18m. The declared value of Z_e is 0.35Ω. Circuit protection at the origin of the installation (consumer unit) is by a BS 1361 5 A fuse.

(a) Establish conformity with BS 7671 requirements.

(b) Establish from BS 7671 Appendix 3 the actual disconnection time.

This is a fixed-equipment circuit; this circuit must disconnect within 0.4s. From Tables I1 and I2 (*IET On-Site Guide*) $R_1 + R_2$ of $1.5mm^2/1.0mm^2$ conductors is $30.2m\Omega/m \times 1.20$. Thus $R_1 + R_2$ of $1.5mm^2/1.0mm^2$ conductors 18m long will be $\dfrac{18 \times 30.2 \times 1.20}{1000} = 0.65\Omega$

and

$Z_s = 0.35 + 0.65$

$= 1.0\Omega$

(a) From Tables B5i and B5ii (*IET On-Site Guide*) maximum measured earth fault loop impedance is $13.68 \times 1.06\Omega$, i.e. 14.5Ω, thus the estimated value of the earth fault loop impedance for this circuit is acceptable.

(b) Actual disconnection time

Prospective earth fault current $= \dfrac{230}{1}$

$= 230A$

From Appendix 3, Table 3A.1, the circuit disconnects in less than 0.1s; we may say that the fuse operates instantaneously.

EXAMPLE 2

A commercial cooker circuit is fed by $16mm^2$ single-core pvc-insulated cable with a $6mm^2$ single-core pvc-insulated protective conductor cable from a BS 88 (Gg) Part 2 type fuseboard (40A fuse) at the origin of the installation; length of cables within pvc conduit is 35m. Assume a tested Z_e value of 0.7.

(a) Establish conformity with BS 7671 requirements regarding the value of Z_s.

(b) Establish from BS 7671 Appendix 3 the actual disconnection time.

(a) This is a fixed-equipment circuit; 5s disconnection time. From Tables I1 and I2 (*IET On-Site Guide*) $R_1 + R_2$ of $16mm^2/6mm^2$ conductors is $4.23m\Omega/m \times 1.20$. Thus $R_1 + R_2$ of $16mm^2/6mm^2$ conductors 15m long will be $\dfrac{35 \times 4.23 \times 1.20}{1000} = 0.178\Omega$

and

$Z_s = 0.7 + 0.178$

$\quad = 0.878\Omega$

From Tables B3i and B3ii (*IET On-Site Guide*) maximum measured earth fault loop impedance is $1.13 \times 1.06\Omega$, i.e. 1.2Ω, thus the estimated value of the earth fault loop impedance for this circuit is acceptable.

(b) Actual disconnection time

Prospective earth fault current $= \dfrac{230}{0.878}$

$\qquad\qquad\qquad\qquad\qquad = 261A$

From Appendix 3, Table 3.B3, the circuit disconnects in 0.6s.

EXAMPLE 3

A cable feeds a single-phase electric pump and lighting point in an external building of a factory, the total length of the twin with earth $4mm^2/2.5mm^2$ pvc-insulated cable is 30m, protection is by a BS 3871, type 3, 30A mcb in a distribution board at the suppliers' intake position. The tested value of Z_s at the intake position is 0.35Ω. This is a fixed-equipment circuit but is in adverse conditions, thus this circuit must disconnect within 0.4s (BS 7671 requirement 411.3.2.2). From Tables I1 and I2 (*IET On-Site Guide*) $R_1 + R_2$ of $4mm^2/2.5mm^2$ conductors is $12.02m\Omega/m \times 1.20$.

Thus $R_1 + R_2$ of $4mm^2/2.5mm^2$ conductors 30m long will be

$$\frac{30 \times 12.02 \times 1.20}{1000} = 0.43\Omega$$

and

$Z_s = 0.35 + 0.43$

$\quad = 0.78\Omega$

From Tables B6 (*IET On-Site Guide*) maximum measured earth fault loop impedance is $0.64 \times 1.06\Omega$, i.e. 0.67Ω, thus the estimated value of the earth fault loop impedance for this circuit is not acceptable. A residual current device in the supply to the external building will be necessary.

EXERCISE 18

Note: Assume that copper conductor cables are used for all questions.

1. A single-phase process heater circuit is wired in pvc trunking and is to employ 6mm² single-core pvc-insulated live conductors and a 2.5mm² protective conductor. The distance from the BS 88 (Gg) Part 2 distribution fuseboard at the main switchgear is 33m, rating of fuse is 40A and ambient temperature is 25°C. The tested Z_e value at the main switchgear is 0.3Ω.

   (a) Estimate the prospective value of Z_s.

   (b) State the maximum permissible measured Z_s value.

2. A single-phase lighting circuit in a commercial premises is wired in pvc conduit employing 1.5mm² live conductors and a 1.5mm² protective conductor. The distance from the BS 1361 distribution fuseboard at the main switchgear is 20m, rating of the fuse is 15A and ambient temperature is 30°C. The tested Z_e value at the main switchgear is 0.45Ω.

   (a) Estimate the prospective value of Z_s.

   (b) State the maximum permissible measured Z_s value.

3. A three-phase electric motor circuit is to be wired in steel trunking and is to employ 4mm² live conductors and the client demands that an independent 2.5mm² protective conductor is used. The distance from the BS 88 (Gm) Part 2, distribution fuseboard at the main switchgear is 10m, and the rating of the fuse is 10A. The testing Z_s value at the distribution board is 0.45Ω, and the ambient temperature is 25°C.

   (a) Estimate the prospective value of Z_s at the motor starter.

   (b) State the maximum permissible measured Z_s value.

4. A 400/230V 50Hz three-phase milling machine is to be wired in pvc trunking and is to employ 6mm² live conductors and the designer specifies that an independent 4mm² protective conductor is to be used. The distance from the BS 88 (Gm) Part 2, distribution fuseboard at the main factory switchgear is 18m, and the rating of the fuse is 50A. The tested Z_s value at the distribution board is 0.4Ω, and the ambient temperature is 20°C.

(a) Estimate the value of Z_s at the machine's starter isolator.

(b) Assuming that the actual value of Z_s is as estimated, what earth fault current will flow in the event of a direct to earth fault at the isolator?

(c) What will be the approximate disconnection time?

5. An earth fault current of 250A occurs in a circuit protected by a BS 88 (Gg) Part 2 32A fuse. The disconnection time will be approximately:

    (a) 0.1s   (b) 0.2s   (c) 0.25s   (d) 3s

6. An earth fault current of 130A occurs in a circuit protected by a BS 3036 30A fuse. The disconnection time will be approximately:

    (a) 0.8s   (b) 0.13s   (c) 1s   (d) 8s

7. An earth fault current of 300A occurs in a circuit protected by a BS 1361 45A fuse. The disconnection time will be approximately:

    (a) 0.18s   (b) 1.8s   (c) 0.3s   (d) 0.9s

Lighting Calculations

14

LIGHTING UNITS AND QUANTITIES

Luminous intensity is the power of light from the source measured in **Candela**.

Illuminance is a measure of the density of luminous flux at a surface measured in **LUX** (lumens per square metre).

Luminous flux is the light emitted by a source and is measured in **LUMENS**.

Luminance is a measure of the light reflected from a surface measured in **CANDELA PER m^2**.

Luminous efficacy is the ratio of the luminous flux emitted by a lamp to the power the lamp consumes, this is measured in **LUMENS PER WATT**.

Quantity	Quantity symbol	Unit	Unit symbol
Luminous intensity	I	Candela	cd
Luminous flux	Φ	Lumens	lm
Illuminance	E	Lux	lx
Luminance	L	Candela per sq m	cd/m^2
Luminous efficacy		Lumens per watt	lm/W

LIGHTING CALCULATIONS USING INVERSE SQUARE LAW

When using the inverse square law the distance used in the measurement is from the light source to a point directly below it.

When a lamp is suspended above a surface, the illuminance at a point below the lamp can be calculated:

$$Illuminance \ E = \frac{(I)candela}{distance^2} \ (ans\, in\, lux)$$

$$= \frac{I}{d^2}$$

EXAMPLE 1

A luminaire producing a luminous intensity of 1500 candela in all directions below the horizontal is suspended 4m above a surface. Calculate the illuminance produced at the surface immediately below the luminaire (Figure 80).

Figure 80 A luminaire

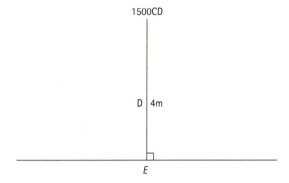

1500CD

D | 4m

E

$$E = \frac{I}{d^2}$$

$$E = \frac{1500}{4^2}$$

$$= 93.75 \ lux$$

EXAMPLE 2

If the luminaire in Example 1 was raised by 1m what would the new illuminance be at the point immediately below the surface?

$E = \dfrac{1500}{(4+1)^2}$

$= \dfrac{1500}{5^2}$

$= 60 \; lux$

LIGHTING CALCULATIONS USING THE COSINE LAW

When using the cosine law the distance used is from the light source measured at an angle to the point at which the lux value is required.

When a lamp is suspended above a horizontal surface, the illuminance E at any point below the surface can be calculated (Figure 81).

Figure 81 Calculate the luminance

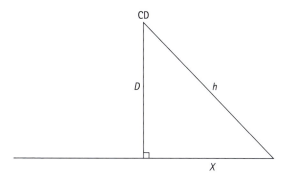

$E = \dfrac{I}{h^2} cos\phi$

To calculate

$h^2 = d^2 + x^2$

To calculate cosø

$Cos\phi = \dfrac{d}{h}$

EXAMPLE 1

A light source producing 1500 candela is suspended 2.2m above a horizontal surface. Calculate the illumination produced on the surface 2.5m away. (Q)

Calculate h^2 using Pythagoras.

$h^2 = d^2 + x^2$

$\quad = 2.2^2 + 2.5^2$

$\quad h^2 = 11.09$

Calculate h

$= \sqrt{h^2}$

$= 3.33$

Figure 82 Calculate the luminance

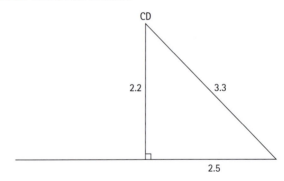

Calculate $\cos\theta$ using Pythagoras

$= \dfrac{d}{h}$

$= \dfrac{2.2}{3.33}$

$= 0.66$

$$E_Q = \frac{1500}{11.09} \times 0.66$$

$= 89.26\ lux$ (Figure 82)

EXAMPLE 2

Two lamps are suspended 10m apart and at a height of 3.5m above a surface (Figure 83). Each lamp emits 350 candelas. Calculate:

(a) the illuminance on the surface midway between the lamps,
(b) the illuminance on the surface immediately below each of the lamps.

Figure 83 Calculate the luminance

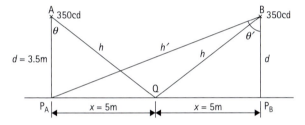

(a) For one lamp, the illuminance at Q is

$$E_Q = \frac{I}{h^2} \cos\Theta$$

$$= \frac{350}{3.5^2 + 5^2} \times \frac{3.5}{\sqrt{(3.5^2 + 5^2)}}$$

$$= \frac{350}{12.25 + 25} \times \frac{3.5}{\sqrt{12.25 + 25}}$$

$$= \frac{350}{37.25} \times \frac{350}{\sqrt{37.25}}$$

$$= 5.388\ lux$$

The illuminance from two lamps is double that due to one lamp, since the conditions for both lamps are identical. Thus total illuminance at Q = 2 × 5.388 = 10.78 lux

(b) At P_A below lamp A, the illuminance due to lamp A is $E_{PA} = \dfrac{I}{d^2}$

$= \dfrac{350}{3.5^2}$

$= 28.57$ lux

In calculating the illuminance at P_A due to lamp B, we have a new distance h', a new distance x', and a new angle Θ' to consider.

$X' = 2x$

$\quad = 10$

$(h')^2 = 3.5^2 + 10^2$

$\quad = 112.25$

$\therefore (h') = 10.59$

$\cos\Theta' = \dfrac{d}{h'}$

$\qquad = \dfrac{3.5}{10.59}$

$\qquad = 0.331$

$\therefore$ illuminance at P_A due to lamp B is $E_{PB} = \dfrac{350}{112.25} \times 0.331$

$= 1.032$

Total illuminance at $P_A = 28.57 + 1.032$

$\qquad\qquad\qquad\quad = 29.61$ lux

and, as the conditions at P_B are the same as those at P_A, this will also be the illuminance below lamp B.

EXERCISE 19

1 A lamp emitting 250 candelas in all directions below the horizontal is fixed 4m above a horizontal surface. Calculate the illuminance at (a) a point P on the surface vertically beneath the lamp, (b) a point Q 3m away from P.

2 Two luminaires illuminate a passageway. The luminaires are 12m apart. Each emits 240 candelas and is 3m above the floor. Calculate the illuminance at a point on the floor midway between the luminaires.

3 Determine the illuminance at a point vertically beneath one of the luminaires in Question 2.

4 An incandescent filament luminaire is suspended 2m above a level work bench. The luminous intensity in all directions below the horizontal is 400 candelas. Calculate the illuminance at a point A on the surface of the bench immediately below the luminaire, and at other bench positions 1m, 2m and 3m from A in a straight line. Show the values on a suitable diagram.

5 Two incandescent filament luminaires are suspended 2m apart and 2.5m above a level work bench. The luminaires give a luminous intensity of 200 candelas in all directions below the horizontal. Calculate the total illuminance at bench level, immediately below each luminaire and midway between them.

6 A work bench is illuminated by a luminaire emitting 350 candelas in all directions below the horizontal and mounted 2.5m above the working surface.

   (a) Calculate the illuminance immediately below the luminaire.

   (b) It is desired to increase the illuminance by 10%. Determine two methods of achieving this, giving calculated values in each case.

7 A lamp emitting 450 candelas in all directions is suspended 3m above the floor. The illuminance on the floor immediately below the lamp is

   (a) 150 lux   (b) 1350 lux   (c) 50 lux   (d) 0.02 lux

8 If the lamp of Question 7 is reduced in height by 0.5m, the illuminance produced immediately below it is

   (a) 72 lux   (b) 36.7 lux   (c) 129 lux   (d) 180 lux

Mechanics

15

MOMENT OF FORCE

The moment of force about a point is found by multiplying together the force and the perpendicular distance between the point and the line of action of the force. Consider an arm attached to a shaft as in Figure 84. The moment acting on the shaft tending to turn it clockwise is 2N × 0.5m = 1Nm

Figure 84 Arm attached to a shaft

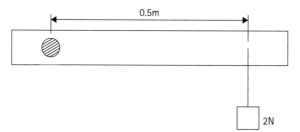

TORQUE

If in Figure 84 a turning effect is applied to the shaft in the opposite direction so that the arm is maintained in a horizontal position, then the *torque* exerted at the shaft is 1Nm.

Consider now an electric motor fitted with a pulley 0.25m in diameter over which a belt passes to drive a machine (Figure 85). If the pull on the tight side of the belt is 60N when the motor is running, then a continuous torque of

$$60N \times \frac{0.25m}{2} = 7.5Nm \text{ is present}$$

Figure 85 An electric motor

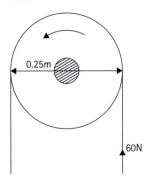

This ignores any pull on the slack side of the belt, and this must usually be taken into account. Thus, if the tension in the slack side of the belt is, say, 10N, then the net torque exerted by the motor is

$$(60-10)N \times \frac{0.25m}{2} = \frac{50 \times 0.25Nm}{2}$$

$$= 6.25Nm$$

In general the torque exerted is $T = (F_1 - F_2) \times r$ Nm

where F_1 is the tension in the tight side, F_2 is the tension in the slack side (in newtons), and r is the pulley radius (in metres).

POWER

$P = 2_{\pi n}T$ watts

where T is the torque in newton metres and n is the speed of the pulley in revolutions per second.

EXAMPLE 1

If the pulley previously considered is running at 16 rev/s, calculate the power output of the motor.

$P = 2_{\pi n}T$

$= 2\pi \times 16 \times 6.25$

$= 629W$

EXAMPLE 2

Calculate the full-load torque of a 3kW motor running at 1200 rev/min.

$1200 \text{ rev/min} = \dfrac{1200}{60} = 20 \text{ rev/s}$

$P = 2_{\pi n}T$

$\therefore 3 \times 1000 = 2\pi \times 20 \times T \text{ (note conversion of kW to W)}$

$\therefore T = \dfrac{3 \times 1000}{2\pi \times 20}$

$= 23.9Nm$

EXAMPLE 3

During a turning operation, a lathe tool exerts a tangential force of 700N on the 100mm diameter workpiece.

(a) Calculate the power involved when the work is rotating at 80 rev/min.
(b) Calculate the current taken by the 230V single-phase a.c. motor, assuming that the lathe gear is 60% efficient, the motor is 75% efficient, and its p.f. is 0.7. The arrangement is shown in Figure 86.

Figure 86 Diagram for Example 3b

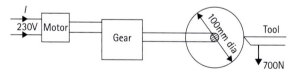

(a) The torque exerted in rotating the work against the tool is

$T = 700N \times 0.05m \text{ (note: radius is 50mm = 0.05m)}$

$= 35Nm$

$$P = 2\pi nT$$

$$= \frac{2\pi \times 80 \times 35}{60} \text{ (note: conversion of rev/min to rev/s)}$$

$$= 293W$$

(b) Motor output $= 293 \times \dfrac{100}{60}$

$$= 488W$$

Motor input $= 488 \times \dfrac{100}{75}$

$$= 650.6W$$

$$P = U \times I \times \text{p.f.}$$

$$\therefore 650.6 = 230 \times I \times 0.7$$

$$\therefore \text{ motor current } I = \frac{650.6}{230 \times 0.7}$$

$$= 4.04A$$

SURFACE SPEED, PULLEY DIAMETER AND SPEED RATIOS

EXAMPLE 1

When turning a piece of low-carbon steel, it should be rotated so that the speed of its surface in relation to the tool is about 0.35m/s. Determine the speed at which a bar 120mm in diameter should be rotated in order to achieve this surface speed. Consider a point on the surface of the steel (Figure 87). In one revolution, this point moves through a distance equal to the circumference of the bar, i.e. distance moved in one revolution $= \pi \times D$

$$= 3.142 \times \frac{120}{1000}$$

$$= 0.377m$$

Numbers of revolutions required for 0.35m $= \dfrac{0.35}{0.377}$

$$= 0.9285$$

$$\therefore \text{ speed of rotation} = 0.928 \text{ rev/s}$$

Figure 87 Rotating steel

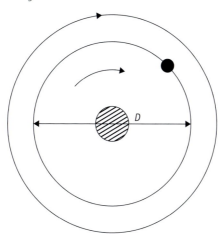

EXAMPLE 2

A machine is driven at 6 rev/s by a belt from a standard motor running at 24 rev/s. The motor is fitted with a 200mm diameter pulley. Find the size of the machine pulley. The speeds at which the pulleys rotate are inversely proportional to their diameters. Thus, if the pulley having a diameter of D_1 rotates at n_1 rev/min and the pulley having a diameter of D_2 rotates at n_2 rev/min (Figure 88), then

$$\frac{n_1}{n_2} = \frac{D_2}{D_1}$$

In this case,

$$\frac{24}{6} = \frac{D_2}{200}$$

$$\therefore D_2 = \frac{200 \times 24}{6}$$

$$= 800\text{mm}$$

Figure 88 Rotating pulley

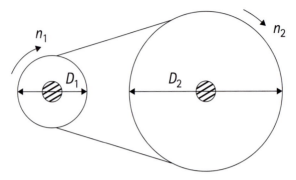

EXERCISE 20

1 A motor drives a machine by means of a belt. The tension in the tight side of the belt is 100N, that in the slack side is 40N, and the pulley is 200mm in diameter. Calculate the total torque exerted by the motor.

2 A test on an induction motor fitted with a prony brake yielded the following results:

Tension in tight side of belt (N)	0	20	30	40	50	60
Tension in slack side of speed (rev/min) belt (N)	0	4	6	8.75	11.5	14.5
Speed (rev/min)	1450	1440	1430	1410	1380	1360

Calculate the torque and power corresponding to each set of readings. Take the pulley radius as being 100mm.

3 A 10kW motor fitted with a 250mm diameter pulley runs at 16 rev/s. Calculate the tension in the tight side of the belt. Ignore any tension in the slack side.

4 A 4kW motor fitted with a 150mm diameter pulley runs at 24 rev/s. The tension in the tight side of the belt may be assumed to be equal to three

times the tension in the slack side. Determine the tension in each side of the belt at full load.

5 Calculate the full-load torque of each of the motors to which the following particulars refer:

	Rated power (kW)	Normal speed (rev/min)
A	10	850
B	2	1475
C	18	750
D	0.25	1480
E	4	1200

6 A motor exerts a torque of 25Nm at 16 rev/s. Assuming that it is 72% efficient, calculate the current it takes from a 440V d.c. supply.

7 A brake test on a small d.c. motor, pulley diameter 75mm, gave the following results:

Net brake tension (N)	0	5	10	15	20	25
Speed (rev/min)	1700	1690	1680	1670	1650	1640
Current (A)	0.8	1.05	1.3	1.68	1.9	2.25
Supply voltage (V)	116	116	116	116	116	116

For each set of values, calculate the power output and the efficiency. Plot a graph of efficiency against power.

8 The chuck of a lathe is driven at 2 rev/s through a gear which is 60% efficient from a 240V d.c. motor. During the turning operation on a 75mm diameter workpiece, the force on the tool is 300N. Calculate the current taken by the motor, assuming its efficiency is 70%.

9 Calculate the speed at the circumference of a 250mm diameter pulley when it is rotating at 11 rev/s.

10 A motor drives a machine through a vee belt. The motor pulley is 120mm in diameter. Calculate the speed at which the belt travels when the motor runs at 24 rev/s.

11 The recommended surface speed for a certain type of grinding wheel is about 20m/s. Determine the speed at which a 250mm diameter wheel must rotate in order to reach this speed.

12 For a certain type of metal, a cutting speed of 0.6m/s is found to be suitable. Calculate the most suitable speed, in revolutions per minute, at which to rotate bars of the metal having the following diameters in order to achieve this surface speed:

    (a) 50mm   (b) 125mm   (c) 150mm   (d) 200mm   (e) 75mm

13 A circular saw is to be driven at 60 rev/s. The motor is a standard one which runs at 1420 rev/min and is fitted with a 200mm diameter pulley. Determine the most suitable size pulley for driving the saw.

14   (a) Calculate the speed of the smaller pulley in Figure 89(a).
       (b) Determine the speed, in rev/min, of the larger pulley in Figure 89(b).

15 Calculate the diameter of the larger pulley in Figure 90(a) and (b).

16 A cutting tool exerts a tangential force of 300N on a steel bar 100mm in diameter which is rotating at 160 rev/min in a lathe. The efficiency of the lathe gearing is 62% and that of the 240V a.c. driving motor is 78%. Calculate the current taken by the motor if its p.f. is 0.6. The pulley on the lathe which takes the drive from the motor is 225mm in diameter and rotates at 600 rev/min. The motor runs at 1420 rev/min. What is the diameter of the motor pulley?

Figure 89 Rotating pulleys

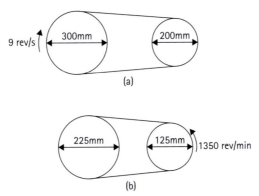

(a)

(b)

Figure 90 Rotating pulleys

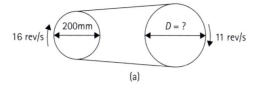

(a)

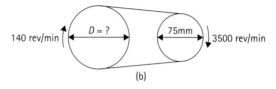

(b)

Miscellaneous Examples 16

D.C. GENERATORS

$$U = E - I_a R_a$$

where

U is the terminal voltage,

E is the generated e.m.f.,

I_a is the armature current, and

R_a is the armature resistance.

Figure 91 Find the shunt field current

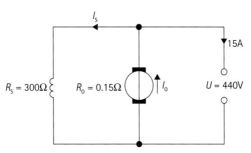

EXAMPLE

Calculate the e.m.f. generated by a shunt generator which is delivering 15A at a terminal voltage of 440V. The armature circuit resistance is 0.15Ω, the resistance of the shunt field is 300Ω, and a voltage drop of 2V occurs at the brushes. The circuit is shown in Figure 91.

To find the shunt field current,

$U = I_s \times R_s$

Where I_s is shunt field current, and R_s is shunt field resistance.

$\therefore 440 = I_s \times 300$

$\therefore I_s = \dfrac{440}{300}$

$\qquad = 1.47A$

Total armature current = 15 + 1.47

$\qquad\qquad\qquad = 16.47A$

Neglecting the voltage drop at the brushes,

$U = E - I_a R_a$

$\therefore 440 = E - 16.47 \times 0.15$

$\qquad\quad = E - 2.47$

$\therefore 440 + 2.47 = E$

$E = 442.47V$

Allowing for the voltage drop at the brushes,

generated e.m.f. = 442.47 + 2

$\qquad\qquad = 444.47$

$\qquad\qquad = 444V$

D.C. MOTORS

$U = E + I_a R_a$ where U is the terminal voltage, E is the back e.m.f., I_a is the armature current, and R_a is the circuit resistance.

EXAMPLE

Calculate the back e.m.f. of a d.c. motor which is taking an armature current of 25A from a 220V supply. The resistance of its armature is 0.2Ω.

$U = E + I_a R_a$

$\therefore 220 = E + 25 \times 0.2$

$\qquad = E + 5.0$

$\therefore 220 - 5 = E$

$\therefore E = 215V$

ALTERNATORS AND SYNCHRONOUS MOTORS

$f = n \times p$

where f is the frequency in hertz, n is the speed in rev/s and p is the number of *pairs* of poles.

EXAMPLE 1

Calculate the number of poles in an alternator which generates 60Hz at a speed of 5 rev/s.

$f = n \times p$

$\therefore 60 = 5 \times p$

$\therefore p = \dfrac{60}{5}$

$\qquad = 12$

$\therefore$ the machine has $2 \times p = 24$ poles

EXAMPLE 2

Calculate the speed at which a four-pole synchronous motor will run from a 50Hz supply.

$f = n \times p$

$\therefore$ 50 = $n \times 2$ (4 poles gives 2 pairs)

$\therefore n = \dfrac{50}{2}$

$\qquad$ = 25 rev/s

INDUCTION MOTORS

percentage slip = $\dfrac{n_s - n_r}{n_s} \times 100\%$

where n_s is the synchronous speed, and n_r is the actual speed of the rotor. The synchronous speed n_s can be determined from the relationship

$f = n_s \times p$

as in the case of the synchronous motor.

EXAMPLE

Calculate the actual speed of a six-pole cage-induction motor operating from a 50Hz supply with 7% slip.

$f = n_s \times p$

$\therefore$ 50 = $n_s \times 3$

$\therefore n_s = \dfrac{50}{3}$

$\qquad$ = 16.7 rev/s

Percentage slip = $\dfrac{n_s - n_r}{n_s} \times 100$

$\therefore 7 = \dfrac{16.7 - n_r}{16.7} \times 100$

$0.07 = \dfrac{16.7 - n_r}{16.7}$

$\therefore$ 0.07 $\times$ 16.7 = 16.7 $- n_r$

$\therefore n_r$ = 16.7 $-$ 0.07 $\times$ 16.7

$\qquad$ = 15.5 rev/s

INSULATION RESISTANCE

The insulation resistance of a cable is *inversely* proportional to its length.

Figure 92 Calculate the insulation resistance

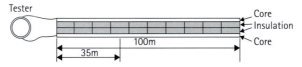

EXAMPLE 1

The insulation resistance measured between the cores of a certain twin cable 100m long is 1000MΩ. Calculate the insulation resistance of 35m of the same cable. The *shorter* length will have a *higher* value of insulation resistance because the path for the leakage current has less cross-sectional area (Figure 92).

Insulation resistance of 100m = 1000MΩ

$$\therefore \text{ insulation resistance of 35m} = 1000 \times \frac{100\,(\text{larger})}{35\,(\text{smaller})}$$
$$= 2857\text{MΩ}$$

EXAMPLE 2

The insulation resistance measured between the cores of a certain twin cable is 850MΩ. Calculate the insulation resistance obtained when two such cables are connected (a) in series, (b) in parallel.

Figure 93 The effect is the same

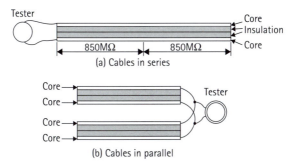

(a) Cables in series

(b) Cables in parallel

It is seen from Figure 93 that the effect in both cases is the same, i.e. to increase the c.s.a. of the leakage-current path through the insulation. The insulation resistance in either case is thus $\frac{850M\Omega}{2} = 425M\Omega$

EXERCISE 21

1 What is meant by the expression 'back e.m.f.' of a d.c. motor? In what way does the back e.m.f. affect the starting of a d.c. motor? A d.c. motor connected to a 460V supply takes an armature current of 120A on a full load. If the armature circuit has a resistance of 0.25Ω, calculate the value of the back e.m.f. at this load.

2 A d.c. machine has an armature resistance of 8Ω. Calculate

  (a) the back e.m.f. when it is operating from a 110V supply and taking an armature current of 2.5A;

  (b) the e.m.f. generated when the machine is running as a generator and delivering 2A at a terminal voltage of 110V. (Neglect the field current.)

3 A d.c. motor connected to a 460V supply has armature resistance of 0.15Ω. Calculate

  (a) the value of the back e.m.f. when the armature current is 120A,

  (b) the value of the armature current when the back e.m.f. is 447.4V.

4 Explain briefly, with the aid of diagrams, the differences between series, shunt and compound d.c. generators. A d.c. shunt generator delivers a current of 96A at 240V. The armature resistance is 0.15Ω, and the field winding has a resistance of 60Ω. Assuming a brush contact drop of 2V, calculate (a) the current in the armature, (b) the generated e.m.f.

5 Calculate the speed at which an eight-pole alternator must be driven in order to generate 50Hz.

6 Calculate the frequency of the voltage generated by a four-pole alternator when it is running at (a) 16 rev/s, (b) 12 rev/s.

7 Determine the speed at which a six-pole synchronous motor will run from the 50Hz mains.

8 The synchronous speed of an induction motor is 750 rev/min. The motor actually runs at 715 rev/min. Calculate the percentage slip.

9 A four-pole induction motor is operating at 24 rev/s from a 50Hz supply. Calculate the percentage slip.

10 A cage-induction motor having six poles operates with a 4.5% slip from a 50Hz supply. Calculate the actual rotor speed.

11 Calculate the full-load torque of a 30kW six-pole 50Hz induction motor, assuming that the slip at full load amounts to 5%.

12 Explain the term 'insulation resistance'. Describe, with wiring diagram, a suitable instrument for measuring insulation resistance. Calculate the insulation resistance of a 100m coil of insulated cable. The insulation resistance of 1km of the same cable is given as 2500MΩ.

13 The insulation resistance of 1000m of two-core cable is 1500MΩ. Calculate the insulation resistance of

    (a) 100m   (b) 200m   (c) 400m   (d) 600m   (e) 800m

and plot a graph showing the relationship between cable length and insulation resistance.

14 Explain the term 'insulation resistance of an installation'. Describe, with connection diagram, the working of an instrument suitable for measuring insulation resistance. Three separate circuits are disconnected from a distribution board and tested for insulation resistance to earth. The respective values are 40MΩ, 60MΩ and 300MΩ. What is the combined insulation resistance to earth?

15 The insulation resistance measured between the cores of a certain twin cable is 950Ω. Calculate the insulation resistance of three identical cables connected in parallel.

16 The resistance of an armature circuit of a motor is 1.2Ω. The current through it is 15A and the terminal voltage is 200V. The generated e.m.f. is

    (a) 218V   (b) 182V   (c) 13.3V   (d) 125V

17 An alternator generates 400Hz at a speed of 500 rev/min. The number of pairs of poles is

    (a) 12   (b) 48   (c) 6   (d) 3

18 The insulation resistance measured between the cores of a cable is 900MΩ for a 500m length. The insulation resistance for 350m of this cable would be:

(a) 1285.7MΩ   (b) 630MΩ   (c) 157.5×10^6 MΩ   (d) 194.4MΩ

Photovoltaic Calculations

For example, if we had six panels each with a V_{oc} of 24V d.c. we would expect a terminal voltage at the inverter of 144V, this would prove that the array was working correctly.

Calculation = 6 panels × 24 volts = 144 volts d.c.

Unfortunately, it is not quite as simple as it seems because the maximum V_{oc} rating given for each PV panel will be the voltage which it would produce under standard test conditions (stc). This involves the manufacturer subjecting the panel to a set level of irradiance at a known temperature. Of course, once the panels have been installed, the levels of irradiance that the panels will be subject to and the temperature of the panels will change constantly.

A calculation is required to check that the array voltage is as expected, on a really bright day the voltage will be close to that which is expected, on an overcast day the voltage will be considerably less.

To carry out a check it is necessary to take into account the temperature coefficient of the panels and the actual temperature of the panels.

Standard test conditions for photovoltaic panel are at a temperature of 25°C and an irradiance of 1000w per m^2.

EXAMPLE 1

A photovoltaic system has one string consisting of eight panels.

A panel data sheet shows a panel to have a V_{oc} of 32.4 volts d.c. and an I_{sc} of 8.59A at the standard test conditions (stc) and a temperature coefficient of 0.27% per degree C.

Voltage check: 8 × 32.5 = 259.2 volts d.c.

This is the voltage expected at 25°C.

The measured temperature at the time if the test was 43°C, this will mean that the voltage will reduce proportionally to the temperature, this test will identify whether or not the panels are operating as they should be.

Calculation to find the expected d.c. voltage at the measured temperature:

Temperature change will be 43°C −25°C = 18°C

Voltage change will be:

$$\frac{18 \times 0.27}{100} = 4.86\%$$

$$\frac{259.2 \times 4.86}{100} = 12.59 \text{ volts}$$

When measured the expected voltage under the conditions described would be:

259.2 − 12.59 = 246.61volts

As you can see the voltage has not changed a great amount which shows us that the temperature does not have a massive influence on the voltage.

The current being generated must also be checked, this requires that the irradiance level is measured and the open circuit current of the array is measured.

EXAMPLE 2

Using the values in Example 1 the I_{sc} is 8.59A at stc which is an irradiance of 1000Wm^2.

An irradiance measurement should be taken, let's assume a measured irradiance of 690Wm^2.

The calculation to find the expected measured short circuit current is

$$\frac{I_{sc} \times \text{measured wm}^2}{1000}$$

$$\frac{8.59 \times 690}{1000} = 5.92\text{A}$$

A far simpler method is to mentally change the irradiance to a decimal and then multiply the I_{sc} by it.

$8.59 \times 0.69 = 5.92A$

EXERCISE 22

1 Calculate the voltage drop in a PV array with an expected open circuit voltage of 263V at stc when measured at a temperature of 48°C. The temperature coefficient for this particular type of panel is 0.35% per degree C.

2 Calculate the expected measured open circuit current for a PV array which has a maximum I_{sc} of 7.5A. The irradiance has been measured as 720Wm².

3 A PV array has a maximum I_{sc} of 8.4A at stc and a measured I_{SC} of 3.8A. Calculate the expected measured irradiance.

Formulae

18

$U = I \times R$ Voltage

$I = \dfrac{U}{R}$ Current

$R = \dfrac{U}{I}$ Resistance

$P = U \times 1$ Power

$P = I^2 R$ Power loss

$I = \dfrac{P}{U}$ Current

$U = \dfrac{P}{I}$ Voltage

$\dfrac{1}{R_1} + \dfrac{1}{R_2} + \dfrac{1}{R_3} = \dfrac{1}{R} \therefore R$ Resistors in parallel

$\dfrac{\pi \times d^2}{4} = C.S.A.$ Area of a circle (mm^2 or m^2)

mm^2 or m^2

$\dfrac{1}{2} base \times height$ Area of a triangle mm^2 or m^2

$\dfrac{1.78 \times 10^{-8} \times L}{CSA \times 10^6} = R$ Resistance of a copper conductor (Ω)

(where c.s.a. is in mm^2)

$\dfrac{2.84 \times 10^{-8} \times L}{CSA \times 10^{-6}} = R$ Resistance of an aluminium conductor (Ω)

(where c.s.a. is mm^2)

Transformer calculation

$$\frac{U_P}{U_S} = \frac{N_P}{N_S} = \frac{I_S}{I_P}$$

Transformer efficiency

$$\frac{power\ out}{power} = per\ unit \times 100\%\ for\ efficiency$$

Work

$W = f \times d\ work\ in\ \dfrac{N}{m} = force\ in\ Newtons \times distance\ in\ mm\ or\ m$

$1 kg = 9.81\ Newtons$

$P = \dfrac{W}{t}\ or\ \dfrac{work\ done\ (Nm)}{Time\ (secs)} = Power\ in\ watts$

$j = W \times t\ or\ Energy\ (joules) = Watts \times time\ in\ seconds$

$E = \dfrac{Input}{Output} \times 100\ Efficiency\ in\%$

Capacitance

Charge of a capacitor is in coulombs $Q = CU$

Total charge of more than one capacitor $Q = Q_1 + Q_2 + Q_3$ etc

$Or\ capacitance\ is\ \dfrac{Q}{U}\ Farads$

$Total\ capacitance\ of\ series\ connected\ \dfrac{1}{C_1} + \dfrac{1}{C_2} + \dfrac{1}{C_3} etc = \dfrac{1}{C_t} = C$

$Total\ capacitance\ of\ parallel\ connected\ C_1 + C_2 + C_3 = C$

Energy stored in a capacitive circuit

$Energy\ W = \dfrac{1}{2}CV^2\ Joules$

Energy stored in an inductive circuit

$Energy\ W = \dfrac{1}{2}LI^2\ Joules\ (where\ L\ is\ henrys)$

THREE-PHASE CALCULATIONS

I_p = Phase current

I_L = Line current

U_L = Line voltage

U_p = Phase voltage

In star *(only one current)*

$$I_P = I_L$$

$$U_P = \frac{U_L}{\sqrt{3}}$$

$$U_L = U_P \sqrt{3}$$

$$P = \sqrt{3} \times U_L \times I_L$$

$$I_L = \frac{P}{\sqrt{3} \times U_L}$$

In circuits with p.f.

$$P = \sqrt{3} \times U_L \times I_L \times Cos\phi$$

$$I_L = \frac{P}{\sqrt{3} \times U_L \times Cos\phi}$$

In delta *(only one voltage)*

$$U_L = U_P$$

$$I_P = \frac{I_L}{\sqrt{3}}$$

$$I_L = I_P \times \sqrt{3}$$

$$P = \sqrt{3} \times U_L \times I_L$$

In circuits with p.f.

$$P = \sqrt{3} \times U_L \times I_L \times Cos\phi$$

$$I_L = \frac{P}{\sqrt{3} \times U_L \times Cos\phi}$$

$$Power\,factor\,cos\,\phi = \frac{True\,power}{Apparent\,power} = \frac{Watts}{Volts \times Amps}$$

PYTHAGORAS–TYPE CALCULATIONS

$Z^2 = R^2 + X^2$ or $Z = \sqrt{R^2 + X^2}$

$R^2 = Z^2 - X^2$ or $R = \sqrt{Z^2 - X^2}$

$R^2 - Z^2 - X^2$ or $R - \sqrt{Z^2 + X^2}$

$X^2 = Z^2 - R^2$ or $X = \sqrt{Z^2 - R^2}$

$kVa^2 = kW^2 = kVar^2$ or $kVa = \sqrt{kW^2 + kVar^2}$

$kW^2 = kVa^2 - kVar^2$ or $kW = \sqrt{kVa^2 - kVar^2}$

$kVar^2 = kVa^2 - kW^2$ or $kVar = \sqrt{kVa^2 - kW^2}$

Capacitive reactance

$X^C = \dfrac{1}{2\pi fC \times 10^{-6}}$ or $\dfrac{1 \times 10^6}{2\pi fC}$

$C = \dfrac{1}{2\pi fX \times 10^{-6}}$ or $\dfrac{1 \times 10^6}{2\pi fX}$

Inductive reactance

$X_L = 2\pi fL$

$L = \dfrac{X_L}{2\pi fX}$

SYNCHRONOUS SPEED AND SLIP CALCULATIONS

N_S is synchronous speed in revs/sec or $\times$ 60 for revs/min

N_R is speed of rotor in revs/sec or $\times$ 60 for revs/min

f is frequency of supply

P is pairs of poles

Unit slip is shown as a decimal

Percentage slip is shown as %

Synchronous speed

$$N_S = \frac{f}{P} \text{ in revs per second} \times 60 \text{ for rpm}$$

Rotor speed

$$\frac{N_S - N_R}{N_S} = \text{unit slip} \times 100 \text{ for \%}$$

CALCULATIONS ASSOCIATED WITH CABLE SELECTION

$$I_t \geq \frac{I_N}{\text{Correction factors}}$$

Cable resistance @ 20 °C

$$R = \frac{r^1 + r^2 \times \text{ length in metres}}{1000}$$

Voltage drop in cable

$$\frac{mV \times \text{Amperes} \times \text{Length}}{1000}$$

Earth fault loop impedance

$$Z_s = Z_e = (R_1 + R_2)$$

Glossary

a.c.	Alternating current
Area	Extent of a surface
BS 7671	British Standard for electrical wiring regulations
Capacitive reactance	The effect on a current flow due to the reactance of a capacitor
Circle	Perfectly round figure
Circuit breaker	A device installed into a circuit to automatically break a circuit in the event of a fault or overload and which can be reset
Circuit	Assembly of electrical equipment which is supplied from the same origin and protected from overcurrent by a protective device
Circumference	Distance around a circle
Conductor	Material used for carrying current
Coulomb	Quantity of electrons
Correction factor	A factor used to allow for different environmental conditions of installed cables
c.s.a.	Cross-sectional area
Current	Flow of electrons
Cycle	Passage of an a.c. waveform through 360°

Cylinder	Solid or hollow, roller-shaped body
d.c.	Direct current
Dimension	Measurement
Earth fault current	The current which flows between the earth conductor and live conductors in a circuit
Earth fault loop impedance	Resistance of the conductors in which the current will flow in the event of an earth fault. This value includes the supply cable, supply transformer and the circuit cable up to the point of the fault
Efficiency	The ratio of output and input power
Energy	The ability to do work
E.M.F.	Electromotive force in volts
Frequency	Number of complete cycles per second of an alternating wave form
Fuse	A device installed in a circuit which melts to break the flow of current in a circuit
Force	Pull of gravity acting on a mass
Hertz	Measurement of frequency
Impedance	Resistance to the flow of current in an a.c. circuit
Impedance triangle	Drawing used to calculate impedance in an a.c. circuit
Internal resistance	Resistance within a cell or cells
I_{SC}	Short circuit current
Kilogram	unit of mass
kW	True power (×1000)
kVA	Apparent power (×1000)

kVAr	Reactive power (×1000)
Load	Object to be moved
Load	The current drawn by electrical equipment connected to an electrical circuit
Mutual induction	Effect of the magnetic field around a conductor on another conductor
Magnetic flux	Quantity of magnetism measured in Webers
Magnetic flux density	The density of flux measured in Webers per metre squared or Tesla
Newton	Pull of gravity (measurement of force)
On-Site Guide	Publication by the IET containing information on electrical installation
Ohm	Unit of resistance
Overload current	An overcurrent flowing in a circuit which is electrically sound
Percentage efficiency	The ratio of input and output power multiplied by 100
Perimeter	Outer edge
Phasor	Drawing used to calculate electrical values
Potential difference	Voltage difference between conductive parts
Power	Energy used doing work
Pressure	Continuous force
Primary winding	Winding of transformer which is connected to a supply
Prospective fault current	The highest current which could flow in a circuit due to a fault
Prospective short circuit current	The maximum current which could flow between live conductors

Protective device	A device inserted into a circuit to protect the cables from overcurrent or fault currents
Rectangle	Four-sided figure with right angles
Resistance	Opposition to the flow of current
Resistor	Component which resists the flow of electricity
Resistivity	Property of a material which affects its ability to conduct
Secondary winding	Winding of transformer which is connected to a load
Self-induction	Effect of a magnetic field in a conductor
Series	Connected end to end
Space factor	Amount of usable space in an enclosure
STC	Standard test conditions
Thermoplastic	Cable insulation which becomes soft when heated and remains flexible when cooled down
Thermosetting	Cable insulation which becomes soft when heated and is rigid when cooled down
Transformer	A device which uses electromagnetism to convert a.c. current from one voltage to another
Transpose	Change order to calculate a value
Triangle	Three-sided object
V_{OC}	Open circuit voltage
Voltage drop	Amount of voltage lost due to a resistance
Volume	Space occupied by a mass

Wattmeter	Instrument used to measure true power
Waveform	The shape of an electrical signal
Work	Energy used moving a load (given in Newton metres or joules)

Answers to Exercises

20

EXERCISE 1

1

Volts V (a.c.)	10	225	230	400	100	25	230	625
Current (A)	0.1	15	0.5	0.4	0.01	500	180	25
Impedance (Ω)	100	15	460	1000	10000	0.05	1.3	25

2

Current (A)	1.92	3.84	18.2	2.38	7.35	4.08	4.17	8.97
Volts V (a.c.)	4.7	7.5	225.7	230	107	228.5	400	235
Impedance (Ω)	2.45	1.95	12.4	96.3	14.56	56	96	26.2

3

Impedance (Ω)	232	850	695.6	0.125	29.85	1050	129	4375
Volts V (a.c.)	176.3	230	400	26.5	0.194	457.8	238	245
Current (A)	0.76	0.27	0.575	212	0.0065	0.436	1.84	0.056

4 98.29Ω
5 1.096A
6 (a) 2.18Ω   (b) 4.49Ω
7 (a) 0.472Ω   (b) 3.83Ω   (c) 0.321Ω   (d) 13Ω   (e) 0.413Ω

8 84.3Ω

9

Volts V (a.c.)	61.1	105	153	193	230
Current (A)	2.3	4.2	6.12	7.35	9.2
Impedance (Ω)	26.56	25	25	26.26	25

10 b

11 c

12 d

EXERCISE 2

1 4.71Ω

2 0.478H

3

Inductance (H)	0.04	0.159	0.12	0.008	0.152
Frequency (Hz)	50	50	48	90	60
Reactance (Ω)	12.57	50	36	4.5	57

4 (a) 40.8Ω (b) 0.13H

5 (a) 16.6A (b) 13.6A

6 (a) 3.77Ω (b) 2.2Ω (c) 0.141Ω (d) 0.11Ω (e) 14.1Ω

7 (a) 0.955H (b) 0.0796H (c) 0.0462H (d) 0.398H (e) 0.0159H

8 398V

9 a

10 c

EXERCISE 3

1 (a) 53Ω (b) 127Ω (c) 79.6Ω (d) 21.2Ω (e) 397Ω (f) 265Ω
 (g) 12.7Ω (h) 33.5Ω (i) 199Ω (j) 42.4Ω

2 (a) 13.3μF (b) 42.4μF (c) 265μF (d) 707μF (e) 88.4μF (f) 199μF
 (g) 70.7μF (h) 7.96μF (i) 106μF (j) 44.2μF

3 346µF
4 6.36µF
5 159V
6 207.6µF
7 364V
8 10A
9 15.2A
10 d
11 a

EXERCISE 4

1

R	15	25	3.64	47.44	4.32	6.32	76.4	0.54
R^2	225	625	13.25	2250	18.7	40	5837	0.735

2

X	29.8	0.68	0.16	0.95	0.4	897	233.7	0.189
X^2	888	0.46	0.026	0.9	0.16	804609	54616	0.036

3 6.71A
4 8.69A
5

R (Ω)	14.5	140	9.63	3.5	57.6	94.8
X (Ω)	22.8	74.6	15.68	34.7	4050	49.6
Z (Ω)	27.02	159	18.4	34.87	4050	107

6 232Ω
7 17.46µF
8 (a) 16.9Ω   (b) 73.3Ω   (c) 71.3Ω
9 0.13H, 115V
10 (a) 28.75Ω   (b) 0.122H   (c) 47.9Ω

11 18.93Ω, 15.04Ω, 0.0479H, 11.5Ω

12 69μF

13 0.318H, 38.9μF, 45.3Hz

14 14.57A

15 (a) 7.47A   (b) 127μF

16 c

17 c

EXERCISE 5

1 50Ω

2 40.1Ω

3 50Ω

4 198Ω

5 46.3Ω

6 231Ω

7 28Ω

8 1.09Ω

9 355Ω

10 751Ω

11 283Ω

12 approx. 500Ω

13

Angleφ	30°	45°	60°	90°	52°24′	26°42′	83°12′	5°36′
sinφ	0.5	0.7071	0.8660	1	0.7923	0.4493	0.9930	0.0976
cosφ	0.866	0.7071	0.5	0	0.6101	0.8934	0.1184	0.9952
tanφ	0.5774	1	1.7321	0	1.2985	0.5029	8.3863	0.0981

14

Angleφ	33°3′	75°21′	17°15′	64°29′	27°56′	41°53′
sinφ	0.5454	0.9675	0.2965	0.9025	0.4384	0.6676
cosφ	0.8382	0.2529	0.9550	0.4308	0.8835	0.7445
tanφ	0.6506	3.8254	0.3105	2.0949	0.5302	0.8967

15

Angleϕ	21°48′	25°48′	65°30′	36°52′	36°52′	50°24′	65°20′	61°36′
sinϕ	0.3714	0.4352	0.91	0.6	0.6	0.7705	0.9088	0.8797
cosϕ	0.9285	0.9003	0.4146	0.8	0.8	0.6374	0.4172	0.4754
tanϕ	0.4000	0.4835	2.1948	0.75	0.75	1.2088	2.1778	1.8505

16

Angleϕ	75°3′	64°16′	5°25′	38°34′	29°38′	72°24	72°23	71°27′
Sinϕ	0.9661	0.9008	0.0946	0.6234	0.4945	0.9532	0.9531	0.9481
cosϕ	0.2582	0.4341	0.9955	0.7819	0.8692	0.3020	0.3026	0.318
tanϕ	3.7346	2.0752	0.0950	0.7973	0.5689	3.152	3.15	2.9814

17 21.3Ω,   20Ω
18 3.95Ω,   6.13Ω
19 31°47′
20 90.9Ω,   78Ω
21 191W,   162VAr
22 32°51′,   129Ω
23 28°57′
24 66.6Ω
25 36.97Ω
26 37.6
27

Phase angleϕ	75°30′	72°30′	65°6′	60°	56°37′	53°7′	45°40′	34°54′
Power factor cosϕ	0.25	0.3	0.421	0.5	0.55	0.6	0.699	0.82

28 (a) 66.5Ω   (b) 1.5A   (c) 0.526   (d) 79.1W
29 (a) 24.78A   (b) 5700VA   (c) 0.86

30 (a) 17Ω   (b) 0.03H   (c) 12.92A   (d) 0.844
31 4Ω, 6.928Ω, 8Ω

EXERCISE 6

1 230V
2 31.1A, 14.1A
3 151V, 44°30′
4 3.63A

EXERCISE 7

1 2.48V
2 1029Ω, 2754Ω
3 7kW, 7.14kVA
4 2130VA
5 179W
6 5.1A
7 2.5A
8 197V

EXERCISE 8

1 0.3 (lag)
2 3.4A, 27°55′ (lag), 0.88 (lag)
3 1.41A (lag)
4 2.78A, 0.86 (lag)
5 3.49A, 0.92 (lag)
6 21.2µF
7 1A
8 b
9 c

EXERCISE 9

1 21.74A (a) 4kW   (b) 3kW
2 131kW, 141kVA
3 6.72kW, 8.97kVA, 356µF
4 11.5kVA, 4.6kW, 7.09kVAr
5 13.8kVA, 6.9kW, 8.6kVAr, 37.39A
6 29.24A
7 124µF, 5.35A
8

Power factor	0.7	0.75	0.8	0.85	0.9	0.95	1.0
Capacitance required µF	1137	1624	2109	2589	3092	3619	4825

9 31.1A, (a) 414µF   (b) 239µF
10 approx. 15µF
11 b
12 a
13 b
14 c

EXERCISE 10

1 (a) 4.62A   (b) 2561W
2 (a)1.63A   (b) 1129W
3 (a) L_1 = 9.2A   (b) L_2 = 17.69A   (c) L_3 = 11.51A
4 (a) 9.58A   (b) 5520W
5 (a) 23.09A   (b) 69.28A
6 (a) L_1 = 20.99A   (b) L_2 = 28.86A   (c) L_3 = 24.93A
7 (a) 7.66A, 7.66A, 5.3kW   (b) 3.33A, 23.09A, 15.9kW
8 (a) 19.21A   (b) 13.3kW
9 (a) 5.17A, 6.2kW   (b) 2.97A, 2.06kW
10 (a) 2.16A, 0.468 lag, 1.5kW   (b) 6.49A, 0.468 lag, 4.5kW
11 (a) 6.64Ω   (b) 20Ω
12 (a) 884µF   (b) 295µF

13 $L_1 - L_2$ = 6.66A $L_2 - L_3$ = 13.33A $L_1 - L_3$ = 12.6A, 14.1kW
14 (a) 6.09kW   (b) 22.6A
15 (a) 7.1kW   (b) 18.86A
16 (a) 17.86A   (b) 37.73A   (c) 26.14kW
17 (a) 17.6kV   (b) in delta V_L = 9.84A, in star V_L and V_P = 433.7A

EXERCISE 11

1 5.29A
2 15.35A
3 17.32A

EXERCISE 12

1 385V, 3.6%, 756W
2 (a) 410V   (b) 1160W
3 12.14V
4 95mm^2
5 467mm^2, 500mm^2, 4.61V
6 5.8V
7 70mm^2
8 25mm^2
9 70mm^2
10 14A
11 (a) 17.32A   (b) 20A   (c) 19.2mV/A/m   (d) 29.35A   (e) 6mm^2
   (f) 3.79V   (g) 25mm
12 (a) (i) 1210A   (ii) 0.25s   (iii) 0.393W   (iv) 585.2A   (v) 3s
   (b) 0.44Ω
13 (a) i) 62.36A   (ii) 63A   (iii) 3.2mV/A/m   (iv) 67.02A
   (v) 16mm^2 (vi) 4.49V
   (b) (i) 0.51Ω (ii) 0.86Ω (Table 41.4)
14 (a) 96.51A   (b) 100A   (c) (i) 97.08A   (ii) 70mm^2 (iii) 1.82V
15 (a) 36.08A   (b) 40A   (c) 42.55A   (d) 2.92 mV/A/m
   (e) 16mm^2 (f) 8.22V

16 (a) 17.32A   (b) 20A   (c) 9.6mV/A/m   (d) 21.3A
    (e) 4mm²   (f) 4.94V   (g) 13.3A (satisfactory)
    (h) table 5C – factor = 260 (satisfactory)

EXERCISE 13

1 40mV
2 45.2Ω
3 99975Ω
4 1.5×10^{-3}Ω
5 9960Ω, 149960Ω, 249990Ω, 40×10^{-3}Ω, 4×10^{-3}Ω
6 c
7 d
8 a
9 b

EXERCISE 14

1 (a) 44.36A   (b) 35.3A   (c) 36.1A   (d) 66A   (e) 79.3A
    (f) 8.05A   (g) 20.9A   (h) 65A
2 84%
3 85.3%, 0.76
4 76.9%, 0.754
5 18.23A
6 d
7 a

EXERCISE 15

1 183.72A. Thermal storage is probably on its own installation, if the shower could be on its own control, then normal 100A consumer unit can be used.
2 152.35A. Propose that the under-sink heaters be on their own consumer unit.

3 230.32A. See paragraph 2 of *IET On-Site Guide*. This is a single-phase supply at present; consultation with the supplier would be essential. Perhaps a poly-phase supply would be available but could incur additional service cable costs.

4 177.72A (approx. 60A per phase)

EXERCISE 16

1 0.448Ω
2 0.538Ω
3 1.02Ω
4 yes
5 (a) yes   (b) yes
6 6.77 (10mm^2)
7 (a) line 25mm^2 (b) c.p.c. 1.5mm^2

EXERCISE 17

1 8.6A
2 (a) 1.26V   (b) 88.2W
3 (a) 45A   (b) 55.6A   (c) 16mm^2 (d) 4.9V
4 (a) 13.04A   (b) 7.2V   (c) 5.86V   (d) 224.14V
5 (a) 75.76A   (b) 1.96V
6 (a) 2.22mV   (b) 25mm^2 (c) 4.05V
7 (a) 60.87A   (b) 63A   (c) 105.5A   (d) 1.45mV/A/m
   (e) 35mm^2 (f) 5.14V
8 (a) 43.5A   (b) 50A   (c) 64.73A   (d) 0.96mV/A/m
   (e) 50mm^2 (f) 4.85V
9 (a) 28.58A   (b) 32A   (c) 34A   (d) 5.95mV/A/m
   (e) 16mm^2 (f) 4V   (g) 32mm conduit
10 (a) 39.13A   (b) 4.47mV/A/m   (c) 40A   (d) 55.47A
   (e) 10mm^2 (f) 229.14V, 227.93V, 226.56V
11 (a) 21.74A   (b) 25A   (c) 25A   (d) 19.2mV/A/m
   (e) 4.0mm^2 (f) 4.18V
12 (a) 27.95A   (b) 11.5V   (c) 30A   (d) 30A
   (e) 8.22mV/A/m (f) 6mm^2 (g) 8.16V

13 (a) 34.78A   (b) 40A   (c) 40A   (d) 6.0mm²
    (e) 4.44V

14 (a) 42.45A   (b) 63.7A   (c) 96.8A   (d) 10mV/A/m
    (e) 50mm²   (f) 3.01V

15 b

16 b

17 a

18 c

19 d

EXERCISE 18

1 (a) 0.715Ω   (b) 1.02Ω

2 (a) 1.0Ω   (b) 4.51Ω

3 (a) 0.594Ω   (b) 6.56Ω

4 (a) 0.566Ω   (b) 0.84Ω   (c) 0.3s

5 c

6 c

7 b

EXERCISE 19

1 (a) 15.6 lux   (b) 8 lux

2 4.77 lux

3 27 lux

4 100 lux, 71.6 lux, 35.4 lux, 17.1 lux

5 47.2 lux, 51.2 lux

6 (a) 56 lux   (b) new lamp of 385 candelas or same lamp, new height 2.38m

7 c

8 a

EXERCISE 20

1 6Nm

2

Torque (Nm)	0	1.6	2.4	3.125	3.85	4.55
Power (W)	0	241.3	359	461	556	648

3 795N
4 531N, 177N
5 (a) 112Nm   (b) 13Nm   (c) 229Nm   (d) 1.61Nm
    (e) 31.8Nm
6 7.94A
7

Po (W)	0	33.2	66.0	98.4	130	161
n (%)	0	27.2	43.7	50.5	58.8	61.7

8 1.4A
9 8.64m/s
10 9.05m/s
11 25.5 rev/s
12 (a) 229 rev/min   (b) 91.7 rev/min   (c) 76.4 rev/min
     (d) 57.3 rev/min   (e) 153 rev/min
13 78.8mm
14 (a) 13.5 rev/s   (b) 750 rev/min
15 (a) 291mm   (b) 181mm
16 3.61A, 95mm

EXERCISE 21

1 430V
2 (a) 90V   (b) 126V
3 (a) 442V   (b) 84A
4 (a) 100A   (b) 257V
5 12.5 rev/s
6 (a) 32hZ   (b) 24Hz
7 16.7 rev/s

8 4.7%
9 4%
10 15.9 rev/s
11 301.5 Nm
12 25 000MΩ
13 (a) 15 000MΩ   (b) 7500MΩ   (c) 3750MΩ   (d) 2500MΩ
    (e) 1875MΩ
14 22.2MΩ
15 317Ω
16 b
17 b
18 b

EXERCISE 22

1 21.17V
2 5.4A
3 450Wm^2

21

General Questions and Answers

Answers on page 224

1 The prospective short circuit current at the origin of the consumer's installation must be taken into account when

    a   Estimating the external earth loop impedance

    b   Calculating the maximum demand

    c   Selecting the system of earthing for the supply

    d   Selecting the type of overcurrent protective device to be installed

2 Design calculations involving conductors and an overcurrent protection device is most likely satisfied when

    a I_b is greater than I_z

    b I_n is greater than I_z

    c I_z is lower than I_n

    d I_n is not less than the design current I_b

3 If the cable is totally enclosed in thermal insulation for a distance of 2m, the rating factor (C_i) will be

    a   0.5

    b   0.55

    c   0.68

    d   0.81

4 An arrow appearing through an electronic symbol means the value of the device is

   a   Preset

   b   Light emitting

   c   Variable

   d   Fixed

5 A 12V battery is connected across a set of resistors in series, values being 60Ω, 30Ω, 100Ω and 45Ω. The current flowing in the circuit is

   a   0.51mA

   b   5.1A

   c   51A

   d   51mA

6 Where motor isolators are remote from the motor they must be

   a   Capable of being locked off

   b   Be seen from the motor area

   c   Painted red

   d   18mm from ground level

7 The purpose of an installation specification is to inform the

   a   Contractor of the client's requirements

   b   Electrician on health and safety matters

   c   Client on how to use the installation

   d   Main contractor on which equipment to use

8 An RCBO is a device which is used as

   a   A voltage reduction sensor

   b   An overload protection device only

   c   Both an overcurrent and residual current protection device

   d   A short circuit protection device only

9 The legal requirement for electrical equipment to be maintained in good order is laid down in the

   a   Management of Health and Safety at Work Regulations 1999

   b   Health and Safety at Work Act 1974

   c   Provision and Use of Work Equipment Regulations 1998

   d   Electricity at Work Regulations 1989

10 MΩ is the abbreviation used for

   a   microhms

   b   milli-ohms

   c   megohms

   d   megaohms

11 Warning signs with a blue background are

   a   Safety signs

   b   Prohibitive signs

   c   Mandatory signs

   d   Explanation signs

12 A length of copper wire has a resistance of 8Ω. What would be the resistance if the length of wire were halved and the cross-sectional area doubled?

   a   2Ω

   b   4Ω

   c   8Ω

   d   16Ω

13 Which of the following does not determine the frequency of the Periodic Inspection and Testing of an installation?

    a   The competence of the tester

    b   Frequency of maintenance

    c   External influences to which the installation is subjected

    d   The type of installation

14 D.C. motors create a constant, stationary, magnetic field in the conductors attached to the

    a   Stator

    b   Yoke

    c   Shaft

    d   Armature

15 The type of fault which occurs when a line conductor comes in contact with a neutral conductor is called

    a   An earth loop fault

    b   A short circuit fault

    c   A catastrophic fault

    d   An overload fault

16 The IET Regulations are designed to provide

    a   Safety from fire, shock and burns

    b   Instruction on electrical equipment

    c   A detailed specification of a system

    d   Instructions for every circumstance in an installation

17 The maximum earth fault loop impedance, permitted by BS 7671: 2008, on a TN system for a ring final circuit protected by a 32A BS 88–3 fuse is a   1.14Ω

    b   0.96Ω

    c   1.92Ω

    d   1.09Ω

18 The minimum depth through which a sheathed cable can pass through a joist is

    a   20mm

    b   30mm

    c   40mm

    d   50mm

19 A black label on a fire extinguisher indicates the extinguisher contains

    a   Foam

    b   Carbon dioxide

    c   Dry powder

    d   Water

20 If two equal wattage lamps were connected in series the voltage drop across each of them would

    a   Be the same

    b   Equal the supply voltage

    c   Be greater across the first lamp

    d   Equal the current flowing

21 A STAR connected system has a line voltage of 1000V, what is the phase voltage?

    a   1732V

    b   577V

   c   1000V

   d   400V

22 Protection against objects greater than 1mm diameter has an IP classification of

   a   IP6X

   b   IP4X

   c   IP3X

   d   IP2X

23 A three-phase cage rotor induction motor is required to operate on a 50Hz supply and run at a speed of approximately 750 rpm. The number of pole pairs required will be

   a   2

   b   4

   c   6

   d   8

24 Instrument test leads should comply with

   a   BS 7671

   b   HSE Guidance Note GS 55

   c   HSE Guidance Note GS 38

   d   BS 2001

25 The insulation resistance of two circuits is of 40MΩ and 36MΩ respectively. When tested together what is the total insulation resistance?

   a   76MΩ

   b   4MΩ

   c   22MΩ

   d   19MΩ

26 Which of the following type of fire extinguisher would you not use on an electrical equipment fire?

    a   Carbon dioxide

    b   Extinguishers with black labels

    c   Foam extinguishers

    d   Dry powder

27 An overload current is

    a   A current arising from an earth fault

    b   Occurring after faulty installation

    c   Due to an open circuit

    d   An overcurrent occurring in a circuit which is electrically sound

28 If the plates of a parallel plate capacitor are increased in area, the capacitance will

    a   Not change

    b   Increase

    c   Decrease

    d   Reduce to zero

29 Electrical equipment users should be

    a   Capable of inspecting equipment for obvious defects

    b   Able to test equipment

    c   An electrically competent person

    d   An electrician

30 If a cable run measured on a drawing with a scale of 1:25 is 65cm, what is the actual length of the cable run?

    a   0.162m

    b   162.5m

c   16.25m

d   1.625m

ANSWERS

1 d	**9** d	**17** b	**25** d
2 d	**10** c	**18** d	**26** c
3 a	**11** c	**19** b	**27** d
4 c	**12** a	**20** a	**28** b
5 d	**13** c	**21** b	**29** a
6 a	**14** b	**22** b	**30** c
7 a	**15** b	**23** b	
8 c	**16** a	**24** c	

Additional Questions and Answers

Answers on page 234

1 ISO 9000 is the standard

   a   To ensure safe working standards

   b   To ensure quality control

   c   For manufacturing cables

   d   For working above 950mm

2 CDM regulations apply to construction work that

   a   Does not last for more than 30 days

   b   Lasts for more than 30 days

   c   Is carried out in a domestic environment

   d   Maintenance of boilers and fires

3 COSHH is an abbreviation for

   a   Control of substances helpful to health

   b   Carrying out some heavy hacking

   c   Control of substances hazardous to health

   d   Coming off site happy and healthy

4 RIDDOR is an abbreviation for

   a   Risk of injuries or dangerous diseases occurring at random

   b   Referring to inspectors doing damage over the road

    c   Referring to injuries or dangerous diseases occurring regularly

    d   Reporting of injuries and diseases and dangerous occurrence regulations

5 A mandatory sign is

    a   Square with a blue background

    b   Round with a blue background

    c   Square with a red background

    d   Round with a red background

6 When working in the electrical industry it is not necessary to have

    a   Changeable and flexible working skills

    b   Flexible working hours

    c   Management skills

    d   Transferable skills

7 An advantage of education and training is

    a   You will reach the top of your pay scale quickly

    b   You will be suitable for a management position

    c   It will enable you to work efficiently and lead your colleagues

    d   You will be better equipped and suited for promotion

8 An HSE inspector can

    a   Provide advice only

    b   Enter a premises by appointment only

    c   Can enter a premises at any time and force work to cease

    d   Issue permits to work

9 After receiving an electric shock a person is found to not have a pulse, the immediate action is to a   Carry out CPR

    b   Carry out mouth-to-mouth resuscitation

    c   Put the person in the recovery position

    d   Go for help

10 In a factory the first-aid room should be in a location that is

    a   Off site in a clean area

    b   Central to the work areas

    c   At the entrance to the factory

    d   Convenient for ambulance access

11 In the event of a death or serious injury on site it should normally be reported within

    a   36 hours

    b   In writing within 24 hours

    c   Immediately

    d   By phone within seven days

12 The Environmental Protection Act

    a   Prohibits all discharges to the environment

    b   Prevents pollution

    c   Controls discharges to the environment

    d   Ensures all drains are cleaned regularly

13 ACOPS is an abbreviation for

    a   Approved control of polluting substances

    b   Approvals corresponding to polluting substances

    c   Approved codes of practice

    d   Another collapsed police station

14 A risk is

    a Anything that can cause harm

    b A chance that somebody may be harmed by a hazard

    c A hazard

    d Working at height

15 After an accident a person is found not to be breathing. The immediate action is to

    a Carry out CPR

    b Carry out mouth-to-mouth resuscitation

    c Put the person in the recovery position

    d Go for help

16 A fire triangle includes Heat, Oxygen and

    a Gas

    b Air

    c Matches

    d Fuel

17 A fire extinguisher with blue indication on it contains

    a Water

    b Foam

    c Carbon Dioxide

    d Dry Powder

18 The most suitable fire extinguisher for tackling a small fire involving electrical equipment is

    a A water fire extinguisher

    b A CO_2 fire extinguisher

c   A foam fire extinguisher

d   A halon gas fire extinguisher

19 When a ladder is used as access to a working platform it must extend above the platform by at least

a   Three rungs

b   1.5 metres

c   1.2 metres

d   Five rungs

20 When an extension ladder is used the ladders must overlap by at least

a   Five rungs

b   1 metre

c   1.2 metres

d   Three rungs

21 The earth for an 11kV to 400 Vstar delta transformer is formed by

a   Connecting the core of the transformer to earth

b   Connecting a primary winding to earth

c   Connecting the primary and secondary star points together

d   Connecting the star point to earth

22 Purely resistive power can be calculated by using the formula

a $P = VI$

b $P = \dfrac{V}{I} \cos\phi$

c $P = VI \sin\phi$

d $P = \dfrac{V}{I}$

23 A Residual Current Device (RCD) will

    a   Operate if a short occurs between phase and neutral

    b   Operate if a small overload occurs

    c   Reconnect if the fault clears

    d   Operate if phase and neutral are out of balance

24 On a TT system an acceptable earth path can be obtained by connecting the main earth terminal to

    a   Underground structural steel work

    b   Gas service pipe

    c   Water service pipe

    d   Telecom service cable

25 A pure capacitor of 100uF is connected to a 230V 50Hz supply. The power dissipated will be

    a   7.2W

    b   32W

    c   2300W

    d   0W

26 The resistance of a material at a constant temperature can be found by use of the formula

    a $R = p/a$

    b $R = \dfrac{pa}{l}$

    c $R = \dfrac{pl}{a}$

    d $R = \dfrac{p}{la}$

27 The force on a conductor placed in a magnetic field can be found by use of the formula

a $\quad F = \dfrac{B}{Il}$

b $\quad F = BIl$

c $\quad F = \dfrac{Bl}{I}$

d $\quad F = \dfrac{BI}{l}$

28 Self-starting single-phase induction motors are referred to as

    a   Two-phase

    b   Dual-phase

    c   Split-phase

    d   Shift-phase

29 A 4 pole three-phase cage induction motor is connected to a 400V 50Hz supply, its speed will be

    a   12.5 rps

    b   25 rps

    c   50 rps

    d   100 rps

30 A large three-phase cage rotor induction motor is most commonly started by means of

    a   A rotor resistance starter

    b   A direct on line starter

    c   A star delta starter

    d   A faceplate starter

31 An isolator in a three-phase four-wire system must disconnect

    a   The three phases and the neutral

    b   The three phases only

    c The three phases, the neutral and the earth

    d The neutral only

32 A component consisting of rolled conductive plates with a dielectric of aluminium oxide is a

    a Polyester capacitor

    b Paper capacitor

    c Air dielectric capacitor

    d Electrolytic capacitor

33 If a single loop of wire is rotated in a magnetic field and its output is taken via slip rings and brushes, its output will be

    a Unidirectional

    b d.c.

    c Intermittent

    d a.c.

34 The starting torque of a three-phase wound rotor induction motor can be increased by

    a Adding resistance to the rotor windings

    b Open circuiting the rotor windings

    c Short circuiting the rotor windings

    d Connecting a capacitor to the rotor windings

35 A workshop measuring 10m × 12m needs to be lit to 500 lux. What would be the required lumen output if the light loss factor was 0.65 and the utilisation factor was 0.75?

    a 123 076 lumens

    b 492 lumens

c 29 250 lumens

d 0.117 lumens

36 Which cable would be most appropriate for wiring the critical signal circuits for a fire alarm system?

a FP200

b LSF

c PVC/SWA

d Flat PVC/PVC

37 A single-phase series motor has similar torque/speed characteristics to a

a Capacitor start motor

b d.c. shunt motor

c d.c. series motor

d Cage rotor motor

38 The effect made use of in a transformer, whereby two coils are in a single magnetic field is known as

a Capacitive reactance

b Mutual conductance

c Mutual inductance

d Mutual impedance

39 The speed of a wound rotor induction motor is determined by

a The number of stator windings

b The voltage of the stator windings

c The inductance of the rotor windings

d The resistance in the rotor windings

40 A three-phase star connected motor is supplied by a 400V 50Hz supply. The current drawn from the supply is 50A per phase and the p.f. is 0.8. The power dissipated is

    a   16kW

    b   20kW

    c   27.7kW

    d   34.6kW

ANSWERS

1 b	**11** b	**21** d	**31** b
2 b	**12** c	**22** a	**32** d
3 c	**13** c	**23** d	**33** d
4 d	**14** b	**24** a	**34** a
5 b	**15** d	**25** d	**35** a
6 c	**16** d	**26** c	**36** a
7 c	**17** d	**27** b	**37** b
8 c	**18** b	**28** c	**38** c
9 d	**19** d	**29** b	**39** d
10 b	**20** d	**30** c	**40** c